Lecture Notes in Mathematics

A collection of informal reports and seminars
Edited by A. Dold, Heidelberg and B. Eckmann, Zürich

240

Adalbert Kerber

Mathematisches Institut der Justus Liebig-Universität
Giessen/Deutschla

Representations of Permutation Groups I

Springer-Verlag
Berlin · Heidelberg · New York 1971

AMS Subject Classifications (1970): 20C30

ISBN 3-540-05693-9 Springer-Verlag Berlin · Heidelberg · New York
ISBN 0-387-05693-9 Springer-Verlag New York · Heidelberg · Berlin

Offsetdruck: Julius Beltz, Hemsbach/Bergstr.

Representations of Permutation Groups

Part I

Representations of Wreath Products and Applications to the Representation Theory of Symmetric and Alternating Groups

Preface

As a contribution to the theory of representations of permutation groups the theory of representations of wreath products of finite groups is discussed in this first part with subsequent applications to the theory of representations of symmetric and alternating groups.

The intention is to give a new description and a further development of the representation theory of symmetric and alternating groups and this will be carried on in the following parts. This seems desirable since following the appearance of the only comprehensive treatment of this theory, namely G.de B.Robinson's book "Representation Theory of the Symmetric Group" (Toronto 1961) a number of papers have been published which continued this work. Moreover some of these papers contain results which allow generalizations which connect this theory more closely with the general representation theory of finite groups.

The representation theory of symmetric and alternating groups is summarized as far as is needed here, while a knowledge of the main results of the general representation theory of finite groups over fields is assumed.

The results of the representation theory of the symmetric group whose proofs are omitted here will be treated in detail in the following parts.

I would express my sincerest thanks to Prof.H. Boerner, Prof. H.K. Farahat, Dr. M.H. Peel and Prof. G. de B.Robinson to whom I am greatly indebted for very helpful discussions and stimulating encouragement.

Adalbert Kerber

Contents

Introduction

The derivation of the representation theory of wreath products provides a nice example of the utility of Clifford's theory of representations of groups with normal divisors.

Applications of this theory to the representation theory of symmetric and alternating groups arise from the fact, that centralizers of elements, normalizers of certain subgroups, Sylow-subgroups as well as defect groups are all direct products of wreath products. Thus for example the theory of the generalized decomposition numbers as well as the theory of symmetrized outer products of irreducible ordinary representations of symmetric groups can be described with the aid of this theory.

On the other hand, representations of wreath products with symmetric groups can be described in detail by using the theory of representations of the symmetric group. But nevertheless this is no vicious circle, since the degrees m of the symmetric factors S_m of the applied wreath products $G \sim S_m$ satisfy $m<n$, if n is the degree of the considered symmetric group S_n. Thus on the contrary these applications provide interesting recursion processes.

Besides this description of representations of wreath products and some of their applications some new results on the modular representation theory of symmetric and alternating groups are cited or proved especially with reference to the theory of decomposition numbers.

It should be mentioned that wreath products are involved in some combinatorial and graph-theoretical devices and the new interest of physicists shouldn't be forgotten (see the references).

The first examples of wreath products can be found in A. Cauchy's "Exercises d'analyse et de physique mathématique" (vol III, 1844), in E. Netto's "Substitutionentheorie und ihre Anwendungen auf die Algebra" and in A. Radzig's dissertation entitled "Die Anwendung des Sylow'schen Satzes auf die symmetrische und die alternirende Gruppe" (1895). Wreath products arise in these cases in connection with the construction of Sylow-subgroups of the considered symmetric group.

The first representation-theoretical consideration of wreath products was given by A. Young, who applied his method of deriving the representation theory of the symmetric group to the so-called hyperoctahedral group in 1930 (Young [1]). In modern notation the hyperoctahedral group is a wreath product $S_2 \sim S_n$. W. Specht considered wreath products of the more general form $G \sim S_n$ (G a finite group) in his dissertation (Specht [1], 1932) and described their ordinary representation theory which he generalized to products of the form $G \sim H$ (G completely reducible, H a subgroup of S_n) in 1933 (Specht [2]).

Such groups had already appeared in papers of A. Loewy (Loewy [1], 1927), A. Scholz (Scholz [1], 1930) and B. Neumann (Neumann [1], 1932). In his paper "Kombinatorische Anzahlbestimmungen für Gruppen, Graphen und chemische Verbindungen" G. Polya suggested for $G \sim H$ the name "G-Kranz um H", of which "G-wreath around H" is a translation. Wreath products $G \sim S_n$ were also considered by O. Ore (Ore [1]) and by R. Frucht (Frucht [1]) in 1942.

Since then numerous papers on these groups as well as on the representation theory of special cases have been published (see the papers of Osima, Puttaswamaiah, Kerber); references to some applications to combinatorics, graph theory and physics may be found at the end of these notes.

Chapter I

Wreath products of groups

In the first section, by way of fixing the notation, we give results concerning the symmetric group S_n ($\simeq \{1\}\sim S_n \simeq S_n\sim\{1\}$) and the alternating group A_n ($\simeq \{1\}\sim A_n \simeq A_n\sim\{1\}$). In the second section general wreath products $G\sim H$ are introduced and in the third section attention is restricted to wreath products $G\sim H$ with $H=S_n$ and G a finite group.

1. Permutation groups

A bijective mapping of a set Ω onto itself is called a <u>permutation of</u> Ω. If a set of permutations of a set Ω together with the composition multiplication is a group, we call this group a <u>permutation group on</u> Ω. The group S_Ω of all the permutations of Ω is called <u>the symmetric group on</u> Ω.

The symmetric groups on two finite sets Ω' and Ω'' of the same order $n = |\Omega'| = |\Omega''|$ are obviously isomorphic. Hence we may denote the symmetric groups of finite sets Ω of order $|\Omega| = n$ by S_n and assume that $\Omega = \{1,\dots,n\}$. The elements of Ω are called <u>symbols</u>.

The order of S_n is

1.1 $$|S_n| = n!$$

as is well known. Subgroups of S_n are called <u>permutation groups of degree</u> n; their elements are called <u>permutations of degree</u> n.

A permutation $\pi \in S_n$ (on $\Omega = \{1,\dots,n\}$) is written down in full by putting the images $\pi(i)$ in a row under the symbols $i \in \Omega$, for example

$$\pi = \begin{pmatrix} 1 & \dots & n \\ \pi(1) & \dots & \pi(n) \end{pmatrix} ,$$

for short:

$$\pi = \begin{pmatrix} i \\ \pi(i) \end{pmatrix} .$$

In accord with the notation $\pi(i)$ for the image of the symbol i under the permutation π, products of permutations have to be read from the right to the left:

$$(\pi'\pi)(i) := \pi'(\pi(i)) .$$

For subsets $\Omega' \subseteq \Omega$ let

$$\pi\Omega' := \{\pi(i) \mid i \in \Omega'\} \subseteq \Omega .$$

A permutation of the form

$$\begin{pmatrix} i_1 & i_2 & \cdots & i_{r-1} & i_r & i_{r+1} & \cdots & i_n \\ i_2 & i_3 & \cdots & i_r & i_1 & i_{r+1} & \cdots & i_n \end{pmatrix}$$

is called <u>cyclic</u> or a <u>cycle</u>. To emphasize the number of symbols which are moved by this cycle, we call it an <u>r-cycle</u>. More briefly we write

$$(i_1 \ldots i_r) ,$$

where the 1-cycles $(i_{r+1}), \ldots, (i_n)$ on the symbols which remain fixed have been omitted.

The identity element of S_n, the permutation which consists only of 1-cycles will be denoted by 1 or by 1_{S_n}.

The following is obviously valid:

1.2 $\qquad (i_1 \ldots i_r) = (i_2 \ldots i_r i_1) = \ldots = (i_r i_1 \ldots i_{r-1}) .$

This means that a cycle which arises from a given one by cylically permuting the symbols describes the same permutation.

2-cycles, i.e. permutations which move exactly two symbols of Ω, are called <u>transpositions</u>.

The <u>order</u> of a cycle, i.e. the order of the generated cyclic subgroup $\langle(i_1\ldots i_r)\rangle \leq S_n$, is equal to its length:

1.3 $$|\langle(i_1\ldots i_r)\rangle| = r.$$

The inverse of this cycle is

1.4 $$(i_1\ldots i_r)^{-1} = (i_r i_{r-1}\ldots i_1)\ .$$

<u>Disjoint</u> cycles (i.e. the sets of really moved symbols are disjoint) describe commuting permutations. Each permutation can be written in <u>cycle-notation</u>, i.e. as a product of pairwise disjoint cycles, which are uniquely determined - as permutations (cf. 1.2) - up to their order of occurence.

Because of

1.5 $$(i_1\ldots i_r) = (i_1 i_r)(i_1 i_{r-1})\ldots(i_1 i_2)$$

each cycle, and therefore every permutation, too, can be written as a product of transpositions. Hence S_n is generated by the transpositions.

Since

1.6 $$(i_j, i_k+1) = (i_k, i_k+1)(i_j i_k)(i_k, i_k+1)$$

S_n is generated even by the transpositions of successive symbols. Another system of generators of S_n is $\{(12),(12\ldots n)\}$, for

1.7 $$(1\ldots n)^r(12)(1\ldots n)^{-r} = (r+1,r+2),\ 0\leq r\leq n-2\ .$$

Hence

1.8 $$S_n = \langle(12),(23),\ldots,(n-1,n)\rangle = \langle(12),(1\ldots n)\rangle\ .$$

(For further results on systems of generators compare Coxeter/ Moser [1], Piccard [1]-[3].)

We would like now to describe the conjugacy classes of S_n. The ordered lengths $\alpha_1, \ldots, \alpha_h$ $(\alpha_j \geq \alpha_{j+1})$ of the cyclic factors of $\pi \in S_n$ (with respect to the cycle-notation of π and including the lengths of 1-cycles) form a <u>partition</u>

1.9 $$P\pi := (\alpha_1, \ldots, \alpha_h) = \alpha$$

of n, i.e. they satisfy

1.10 $$\sum_i \alpha_i = n,\ \alpha_i \in \mathbb{N}\ ,\ \alpha_j \geq \alpha_{j+1}\ (1 \leq j \leq n-1)\ .$$

$P\pi$ is called <u>the partition of</u> π.

If α is the partition of π, then we obtain from 1.3 that the order of the generated subgroup $\langle\pi\rangle \leq S_n$ is equal to the least common multiple of the elements α_i of α:

<u>1.11</u> $$P\pi = (\alpha_1, \ldots, \alpha_h) \Rightarrow |\langle\pi\rangle| = \text{lcm}\ \alpha_i\ .$$

Thus we have a criterion whether π is <u>p-regular</u>, <u>p-singular</u> or a <u>p-element</u> with respect to a prime number p (i.e. whether $p \nmid |\langle\pi\rangle|$, $p \mid |\langle\pi\rangle|$ or $|\langle\pi\rangle|$ is a power of p):

<u>1.12</u> <u>If $P\pi = (\alpha_1, \ldots, \alpha_h)$:</u>

<u>(i) π is p-regular $\Leftrightarrow$ $\forall i\colon p \nmid \alpha_i$,</u>

<u>(ii) π is p-singular $\Leftrightarrow$ $\exists i\colon p \mid \alpha_i$,</u>

<u>(iii) π is a p-element $\Leftrightarrow$ $\forall i\colon \alpha_i$ is a power of p .</u>

Now we wish to show that each subset of S_n consisting of the permutations with a certain partition α forms a conjugacy class of S_n. To prove this we notice first that

1.13 $$\pi\pi'\pi^{-1} = \begin{pmatrix} i \\ \pi(i) \end{pmatrix} \begin{pmatrix} i \\ \pi'(i) \end{pmatrix} \begin{pmatrix} \pi(i) \\ i \end{pmatrix} = \begin{pmatrix} \pi(i) \\ \pi\pi'(i) \end{pmatrix} .$$

This means, that we obtain $\pi\pi'\pi^{-1}$ from π' by an application of π to the symbols in the cycle-notation of π', e.g.

$$(123)(35)(123)^{-1} = (15) .$$

This application of π obviously doesn't disturb the lengths of the cyclic factors of π' so that we have

$$P\pi\pi'\pi^{-1} = P\pi' .$$

Hence a conjugacy class consists of permutations of equal partitions.

On the other hand, if we are given $\pi'' \in S_n$ with $P\pi'' = P\pi'$, there obviously exist permutations π which fulfil $\pi\pi'\pi^{-1} = \pi''$, namely all the permutations π which map the symbols of π' onto the symbols of π'' as described above. Therefore denoting by "$\sim$" that the two permutations are conjugates we have

1.14 $$\pi' \sim \pi'' \Leftrightarrow P\pi' = P\pi'' .$$

Because of 1.4 we have $P\pi = P\pi^{-1}$ so that every permutation in S_n is a conjugate of its inverse. Groups in which every element is a conjugate of its inverse are called <u>ambivalent</u>. Hence:

1.15 <u>S_n is ambivalent.</u>

The use of a second notation facilitates the calculation of the order of a conjugacy class. For $i = 1,\dots,n$ let a_i be the number of elements α_j of $\alpha = P\pi$ with $\alpha_j = i$, i.e. the number of i-cycles among the cyclic factors of π. This establishes a one-to-one correspondence between the partitions $P\pi = \alpha$ and the $(1\times n)$-matrices

1.16 $\quad T\pi := (a_1,\ldots,a_n) = a\ ,\ 0\leq a_i \in \mathbb{Z}\ ,\ \sum_i ia_i = n\ .$

$T\pi$ is called <u>the type of</u> π.

Hence we may describe the conjugacy classes of S_n with the aid of such n-tupels, too: Let $C^\alpha = C^a$ denote the class of the permutations of S_n consisting of h cycles of the lengths $\alpha_1, \ldots, \alpha_h$ respectively consisting of a_i i-cycles for $i = 1,\ldots,n$.

Because of 1.13 and 1.2, for π' and π'' out of C^a we have exactly

1.17 $$\prod_i i^{a_i} a_i!$$

permutations π which satisfy $\pi\pi'\pi^{-1} = \pi''$. Hence 1.17 is the order of the centralizer $C_{S_n}(\pi')$ of π' in S_n and we have:

1.18 $\quad |C^\alpha = C^a| = n!/(\prod_i i^{a_i} a_i!)\ ;\ |C_{S_n}(\pi'\in C^a)| = \prod_i i^{a_i} a_i!\ .$

As an example we consider S_3 (for $n\geq 3$, S_n is not abelian):

$$S_3 = \{1,(12),(13),(23),(123),(132)\}\ .$$

This group consists of 3 conjugacy classes corresponding to the partitions (3), $(2,1)$ and $(1,1,1) =: (1^3)$ respectively to the types $(0,0,1)$, $(1,1,0)$ and $(3,0,0)$. These classes are

$$C^{(1^3)} = C^{(3,0,0)} = \{1\}\ ,$$
$$C^{(2,1)} = C^{(1,1,0)} = \{(12),(13),(23)\}\ ,$$
$$C^{(3)} = C^{(0,0,1)} = \{(123),(132)\}\ .$$

We shall consider the group-theoretical structure of the centralizers later on.

Other subgroups of importance are the <u>alternating groups</u> A_n. $A_n \leq S_n$ consists of the permutations $\pi \in S_n$, which don't change the sign of the <u>difference-product</u>

$$\Delta_n := \prod_{1 \leq i < j \leq n} (j-i) ,$$

i.e.

1.19 $$A_n := \{\pi \in S_n \mid \pi\Delta_n := \prod_{1 \leq i < j \leq n} (\pi(j)-\pi(i)) = \Delta_n\} .$$

(It is easy to see that $\pi\Delta_n = \pm\Delta_n$). Recall that the empty difference-product Δ_1 is 1 by convention so that

$$A_1 = A_2 = \{1\} .$$

Permutations $\pi \in S_n \backslash A_n$, i.e. permutations satisfying $\pi\Delta_n = -\Delta_n$, are called <u>odd permutations</u>, the elements of A_n are called <u>even permutations</u>.

It is obvious, that the product of two even permutations is even, hence A_n is a subgroup. Thus every permutation group $P \leq S_n$ contains a subgroup P^+ consisting of its even elements:

1.20 $$P^+ := P \cap A_n .$$

Depending on π, left cosets πP^+ consist either of even or of odd permutations. Since the left coset P^+ includes all the even permutations we obtain for the index of P^+ in P:

<u>1.21</u> $$|P : P^+| \leq 2 .$$

Hence P^+ is a normal divisor of P in any case: $P^+ \trianglelefteq P$.

For $n \geq 2$ S_n contains the transposition (12), which is odd, thus we obtain as a special case of 1.21:

1.22 $$|A_n| = n!/2, \ \forall \ n \geq 2.$$

Since 1.5 is valid, a cycle is even if and only if its length r is odd. E. g.

$$A_3 = \{1,(123),(132)\} \triangleleft S_3 \ .$$

Hence an investigation of the partition or of the type of π allows one to decide whether π belongs to A_n or to $S_n \backslash A_n$. Hence the conjugacy classes of S_n - let's call them <u>S_n-classes</u> - belong to A_n or to $S_n \backslash A_n$ in full. The question arises, which of the S_n-classes split into <u>A_n-classes</u>.

The centralizer of an even permutation π in A_n is

$$C_{A_n}(\pi) = C_{S_n}(\pi)^+ = C_{S_n}(\pi) \cap A_n \ .$$

Hence we obtain from 1.21:

$$\forall \ \pi \in A_n\text{: either } \ C_{A_n}(\pi) = C_{S_n}(\pi) \quad \text{or} \quad |C_{S_n}(\pi):C_{A_n}(\pi)| = 2.$$

Thus we have for the orders of the conjugacy classes (which are equal to the index of the centralizer of each of their elements):

$$\forall \ \pi \in A_n\text{: either } |C^S(\pi)| = 2|C^A(\pi)| \quad \text{or } C^S(\pi) = C^A(\pi) \ .$$

Hence the S_n-class of $\pi \in A_n$ splits - and then into two A_n-classes of the same order - if and only if

$$C_{A_n}(\pi) = C_{S_n}(\pi) \ .$$

We prove a little bit more:

1.23 <u>For n>1 exactly those S_n-classes split (and then into two classes of equal order), the centralizers of whose elements π fulfil</u>

$$C_{S_n}(\pi) = C_{A_n}(\pi) .$$

<u>These are exactly the classes the partitions of whose elements are of the form $\alpha = (\alpha_1,\dots,\alpha_n)$ with pairwise different and odd elements α_i.</u>

Proof: It remains to prove the second part of the statement.

a) A permutation π commutes with each of its cyclic factors. Hence if the class of π splits, π cannot have an odd cyclic factor (of even length).

Analogously we see that π cannot have two cyclic factors $(i_1\dots i_r)$ and $(i_1'\dots i_r')$ of the same odd length r, for this would imply the existence of an odd permutation in the centralizer:

$$S_n \backslash A_n \ni (i_1 i_1')\dots(i_r i_r') \in C_{S_n}(\pi) .$$

It follows that at most those classes split whose permutation α has pairwise different and odd elements α_i.

b) That these classes really split we derive from the odd order $\prod_i \alpha_i$ of the centralizers of their elements which implies

$$C_{S_n}(\pi)^+ = C_{S_n}(\pi) .$$

q.e.d.

If the S_n-class $C^\alpha = C^a$ splits we shall denote by $C^{\alpha\pm} = C^{a\pm}$ the two A_n-classes and fix $C^{\alpha+} = C^{a+}$ by

1.24 $$(1\ldots\alpha_1)(\alpha_1+1\ldots\alpha_1+\alpha_2)\ldots(\ldots n) \in C^{\alpha+} = C^{a+} .$$

Though

$$C^{(3)+} = \{(123)\},\ C^{(3)-} = \{(132)\},$$

it is not true in general that $C^{\alpha-}$ consists of the inverses of the elements of $C^{\alpha+}$. E.g.

$$(25)(34)(12345)(25)(34) = (15432) = (12345)^{-1},$$

so that (12345) as well as $(12345)^{-1}$ belong to $C^{(5)+} \subset A_5$.

We prove that (Berggren [1]):

1.25 <u>$A_1 = A_2 = \{1\}$, A_5, A_6, A_{10} and A_{14} are the only ambivalent alternating groups.</u>

Proof: Because of 1.15 and 1.23 we need only consider A_n-classes $C^{\alpha\pm}$ of partitions $\alpha = (\alpha_1,\ldots,\alpha_h)$ with pairwise different and odd summands α_i. Each of the other A_n-classes contains with an element its inverse.

Let

$$\pi = (i_1\ldots i_r)\ldots(j_1\ldots j_s)$$

be a product of disjoint cycles of odd lengths. We can form

$$\rho := (i_2 i_r)(i_3 i_{r-1})\ldots(j_2 j_s)(j_3 j_{s-1})\ldots ,$$

the <u>standard-conjugator</u> of π which satisfies

$$\rho\pi\rho^{-1} = \pi^{-1} . \qquad (1)$$

We notice, that ρ is an odd permutation if and only if the number of cyclic factors of π whose length is congruent 3 modulo 4 is odd.

a) If the standard-conjugator ρ for such an element π out of a splitting class is odd, then the considered alternating group A_n cannot be ambivalent. For if $\sigma \in A_n$ and $\sigma\pi\sigma^{-1} = \pi^{-1}$ equation (1) would imply $(\sigma^{-1}\rho)\pi(\sigma^{-1}\rho)^{-1} = \pi$, which is in contradiction to $C_{S_n}(\pi) = C_{A_n}(\pi)$ since $\sigma^{-1}\rho \in S_n \backslash A_n$.

b) It remains to show that exactly for the natural numbers $n \notin \{1,2,5,6,10,14\}$ there are partitions with pairwise different and odd summands α_i and so that the number of α_i satisfying $\alpha_i \equiv 3\ (4)$ is odd.

(i)The only partitions α of the $n \in \{1,2,5,6,10,14\}$ with pairwise different and odd elements α_i are as follows:

<u>n = 1</u>: (1); <u>n = 2</u>: $\emptyset$; <u>n = 5</u>: (5); <u>n = 6</u>: (5,1); <u>n = 10</u>: (9,1), (7,3); <u>n = 14</u>: (13,1), (11,3), (9,5),

in each of which the number of elements congruent 3 modulo 4 is 0 or 2 and therfore even. Hence the standard-conjugator is even in every case such that these alternating groups are ambivalent.

(ii)Let us now look at the $n \notin \{1,2,5,6,10,14\}$.

We distinguish the natural numbers n with respect to their residue classes modulo 4.

1. <u>$n = 4k,\ k \in \mathbb{N}$</u>: For n = 4 we have the partition (3,1) and

for k>1 the partition (4k - 3,3) with an odd number of elements congruent 3 modulo 4.

2. n = 4k + 1: Because of the ambivalency of A_1 and A_5 we assume $k \geq 2$. (4k - 3,3,1) fulfils the condition.

3. n = 4k + 2, $k \geq 4$ (A_2, A_6, A_{10}, A_{14} are ambivalent): (4(k-1) - 3, 5,3,1).

4. n = 4k + 3: (n) = (4k + 3) has one element, and this one is congruent 3 modulo 4.

Thus each of the alternating groups A_n with $n \notin \{1,2,5,6,10,14\}$ possesses a conjugacy class which does not contain the inverse of each of its elements and hence these alternating groups are not ambivalent.

q.e.d.

From the proof of 1.25 we get:

1.26 The two A_n-classes $C^{\alpha\pm}$ into which the S_n-class to the partition $\alpha = (\alpha_1,\ldots,\alpha_h)$ with pairwise different and odd elements α_i splits (suppose n>1) are ambivalent if and only if the number of elements α_i of α with $\alpha_i \equiv 3\ (4)$ is even.

We shall come back to the centralizers of elements when we have said something about wreath products, since they are direct products of certain wreath products.

To conclude this section let us look at the double cosets of der certain subgroups which will be of use later on.

A. J. Coleman has pointed to these double cosets (Coleman [1]) as elucidating the group-theoretical background of the apparently purely combinatorial introduction to the representation theory of the symmetric group (cf. section 4).

For a partition $\alpha=(\alpha_1,\dots,\alpha_h)$ of n, let $\Omega_1^\alpha, \dots, \Omega_h^\alpha$ be pairwise disjoint subsets of orders $|\Omega_i^\alpha| = \alpha_i$ of the set $\Omega = \{1,\dots,n\}$ of symbols on which S_n acts.

Subgroups of the form

1.27 $$S_\alpha := S_{\alpha_1} \times \dots \times S_{\alpha_h} = \underset{i}{\times} S_{\alpha_i} \leq S_n$$

(S_{α_i} the subgroup of the $\alpha_i!$ permutations fixing the symbols out of $\Omega\backslash\Omega_i^\alpha$, $1\leq i\leq h$) are called <u>Young-subgroups</u> in honour of A. Young (1873-1940) to whom we are indebted for the theory of representations of the symmetric group in which such subgroups play an important role (cf. section 4).

If we are given two partitions, say $\alpha = (\alpha_1,\dots,\alpha_h)$ and $\beta = (\beta_1,\dots,\beta_k)$, of n and two Young-subgroups $S_\alpha = \times S_{\alpha_i}$ and $S_\beta = \times S_{\beta_j}$, π and π^* out of S_n, we want to show the following (Coleman [1]):

<u>1.28</u> $$\pi \in S_\alpha\pi^*S_\beta \Leftrightarrow \forall\, i,j: |\Omega_i^\alpha \cap \pi\Omega_j^\beta| = |\Omega_i^\alpha \cap \pi^*\Omega_j^\beta| .$$

Proof:

(i) If $\pi = \pi'\pi^*\pi''$, $\pi' \in S_\alpha$, $\pi'' \in S_\beta$, we have $\pi\Omega_j^\beta = \pi'\pi^*\Omega_j^\beta$, $\forall j$.

$$\Rightarrow \Omega_i^\alpha \cap \pi\Omega_j^\beta = \pi'(\Omega_i^\alpha \cap \pi^*\Omega_j^\beta),\ \forall\, i,j.$$

$$\Rightarrow |\Omega_i^\alpha \cap \pi\Omega_j^\beta| = |\Omega_i^\alpha \cap \pi^*\Omega_j^\beta| ,\ \forall\, i,j.$$

(ii) $|\Omega_i^\alpha \cap \pi\Omega_j^\beta| = |\Omega_i^\alpha \cap \pi^*\Omega_j^\beta|$, $\forall$ i,j, implies that the subsets $\Omega_i^\alpha \cap \pi\Omega_j^\beta$ as well as the subsets $\Omega_i^\alpha \cap \pi^*\Omega_j^\beta$ form for fixed j two complete dissections of Ω_j^β into pairwise disjoint subsets which can be collected into pairs $(\Omega_i^\alpha \cap \pi\Omega_j^\beta, \Omega_i^\alpha \cap \pi^*\Omega_j^\beta)$ of subsets of equal order.

Hence for each i there is a $\pi_i' \in S_{\alpha_i}$ satisfying

$$\pi_i'(\Omega_i^\alpha \cap \pi\Omega_j^\beta) = \Omega_i^\alpha \cap \pi^*\Omega_j^\beta \ , \ \forall \ j.$$

Multiplying these π_i' together we have for the resulting $\pi' := \pi_1' \dots \pi_h' \in S_\alpha$:

$$\pi'\pi\Omega_j^\beta = \pi^*\Omega_j^\beta \ , \ \forall \ j.$$

Thus there is a $\pi'' \in S_\beta$ such that $\pi = \pi'^{-1}\pi^*\pi''$ as stated.

q.e.d.

Hence we have a 1-1-correspondence between the set of double cosets $S_\alpha\pi S_\beta$ of S_α and S_β and the system

1.29 $$\{z_{ij} := |\Omega_i^\alpha \cap \pi\Omega_j^\beta|\}$$

of rational integers satisfying

1.30 $$\sum_{i=1}^{h} z_{ij} = \beta_j, \ \sum_{j=1}^{k} z_{ij} = \alpha_i, \ 0 \leq z_{ij} \in \mathbb{Z} \ .$$

We notice that

1.31 $$S_\alpha \cap \pi S_\beta \pi^{-1} = \times_{i,j} S_{z_{ij}} \leq S_n \ ,$$

if $S_{z_{ij}}$ is the subgroup of the $z_{ij}!$ permutations of S_n fixing the elements of $\Omega \setminus (\Omega_i^\alpha \cap \pi\Omega_j^\beta)$, $S_0 := \{1\} \leq S_n$.

The number of such systems $\{z_{ij}\}$ which fulfil 1.30 is obviously equal to the coefficient of

$$x^{\alpha}y^{\beta} := x_1^{\alpha_1}\cdots x_h^{\alpha_h}y_1^{\beta_1}\cdots y_k^{\beta_k}$$

in

$$\prod_{i,j=1}^{h,k}(1-x_iy_j)^{-1} := \prod_{i,j}\left(\sum_{\nu=0}^{\infty}x_i^{\nu}y_j^{\nu}\right)$$

(x_i, y_j independent indeterminates); from which it follows that:

<u>1.32 The number of distinct double cosets $S_\alpha\pi S_\beta$ of S_α and S_β in S_n is equal to the coefficient of $x^\alpha y^\beta$ in $\prod(1-x_iy_j)^{-1}$.</u>

We notice that during these group-theoretical considerations the polynomials $\prod(1-x_iy_j)^{-1}$ make their appearance, which play an important role in the character theory of the symmetric group and in the theory of the so-called S-functions (cf. Littlewood [2], 5.2, 6.4).

Special double cosets $S_\alpha\pi S_\beta$ are those corresponding to solutions $\{z_{ij}\}$ of 1.30 which satisfy $0\leq z_{ij}\leq 1$, i.e. (cf. 1.31)

$$S_\alpha \cap \pi S_\beta\pi^{-1} = \{1\}.$$

For their number we get from 1.32:

<u>1.33 The number of double cosets $S_\alpha\pi S_\beta$ with the property $S_\alpha \cap \pi S_\beta\pi^{-1} = \{1\}$ is equal to the coefficient of $x^\alpha y^\beta$ in $\prod(1+x_iy_j)$.</u>

Concluding this section we consider certain pairs (α,β) of partitions with respect to this number of certain double cosets.

A partition $\alpha = (\alpha_1,\dots,\alpha_h)$ of n can be illustrated by a Young-diagram $[\alpha]$ consisting of n nodes in h rows, with α_i nodes in the i-th row and all the h rows starting in the same column. E.g. the partition $(3,2,1,1) =: (3,2,1^2)$ can be illustrated by

$$[3,2,1^2]: \begin{array}{lll} \cdot & \cdot & \cdot \\ \cdot & \cdot & \\ \cdot & & \\ \cdot & & \end{array}$$

On account of the condition $\alpha_j \geq \alpha_{j+1}$ $(1 \leq j \leq n-1)$ it makes sense to speak of columns of the diagram $[\alpha]$, and of the length α_i' of the i-th column. Hence to the partition α there corresponds the partition

1.34 $$\alpha' := (\alpha_1',\dots,\alpha_{h'=\alpha_1}') \ , \ \alpha_i' := \sum_{j:\alpha_j \geq i} 1$$

α' is called the associated partition of α, $[\alpha']$ the Young-diagram associated with $[\alpha]$ which arises by interchanging the rows and columns, e.g.

$$[3,2,1^{2\prime}] = [4,2,1]: \begin{array}{llll} \cdot & \cdot & \cdot & \cdot \\ \cdot & \cdot & & \\ \cdot & & & \end{array}$$

Partitions α and Young-diagrams $[\alpha]$ with the property $\alpha = \alpha'$ resp. $[\alpha] = [\alpha']$ are called selfassociated, and it will appear in section 4, that the following lemma is crucial:

1.35 Young-subgroups S_α and $S_{\alpha'}$ possess exactly one double coset $S_\alpha \pi S_{\alpha'}$ with the property that $S_\alpha \cap \pi S_{\alpha'} \pi^{-1} = \{1\}$.

Proof: We want to use 1.33 and hence we prove by induction with respect to h, that

$$\{\text{coeff. of } x^{\alpha}y^{\alpha'} \text{ in } \prod_{i,j=1}^{h,h'} (1+x_iy_j)\} = 1 .$$

(i) h=1: Obviously

$$\{\text{coeff. of } x_1^{\alpha_1}\prod_{i=1}^{\alpha_1} y_i \text{ in } \prod_{j=1}^{\alpha_1} (1+x_1y_j)\} = 1 .$$

(ii) If we assume that $\alpha_h = r$, the induction hypothesis is

$$\{\text{coeff. of } (\prod_{i=1}^{h-1} x_i^{\alpha_i})y_1^{\alpha_1'-1} \ldots y_r^{\alpha_r'-1} y_{r+1}^{\alpha_{r+1}'} \ldots y_{h'}^{\alpha_{h'}'} \text{ in } \prod_{i,j}^{h-1,h'}(1+x_iy_j)\}=1 .$$

Multiplying both sides with $\prod_{j=1}^{r} x_hy_j$ gives

$$\{\text{coeff. of } x^{\alpha}y^{\alpha'} \text{ in } \prod_{i,j=1}^{h-1,h'} (1+x_iy_j)\prod_{j=1}^{r}x_hy_j\} = 1 .$$

Since $\alpha_h=r$ the coefficient of $x^{\alpha}y^{\alpha'}$ is the same in

$$\prod_{i,j=1}^{h-1,h'} (1+x_iy_j)\prod_{j=1}^{r}x_hy_j$$

as in

$$\prod_{i,j=1}^{h-1,h'} (1+x_iy_j)\prod_{j=1}^{r}(1+x_hy_j)$$

as well as (since $\alpha_j' \leq h-1$ for $j>r$) in

$$\prod_{i,j=1}^{h-1,h'} (1+x_iy_j)\prod_{j=1}^{h'=\alpha_1}(1+x_hy_j) = \prod_{i,j=1}^{h,h'}(1+x_iy_j) .$$

This together with 1.33 yields the statement.

q.e.d.

The double coset $S_{\alpha}\pi S_{\alpha'}$ with $S_{\alpha} \cap \pi S_{\alpha'}\pi^{-1} = \{1\}$ may be illustrated by a so-called <u>Young-tableau</u> which arises from the Young-diagram $[\alpha]$ by replacing the nodes by the symbols $1,\ldots,n$ of Ω. E.g.

3 5 1
4 6
2
7

is a tableau with diagram $[3,2,1^2]$.

If Ω_i^α, $\Omega_j^{\alpha'}$ are the sets of symbols in the i-th row, j-th column of the tableau T^α, the elements of the Young subgroups $S_\alpha = \times S_{\alpha_i}$ resp. $S_{\alpha'} = \times S_{\alpha'_j}$ (S_{α_i} resp. $S_{\alpha'_j}$ fixing the elements of $\Omega\backslash\Omega_i^\alpha$ resp. $\Omega\backslash\Omega_j^{\alpha'}$) are called <u>horizontal permutations</u> resp. <u>vertical permutations</u> of T^α. The <u>groups of</u> all the <u>horizontal</u> resp. all the <u>vertical permutations</u> of T^α are indicated by

1.36 $$H^\alpha \quad \text{resp.} \quad V^\alpha .$$

They satisfy

1.37 $$H^\alpha \cap V^\alpha = \{1\} .$$

Therefore the tableau T^α illustrates the double coset $S_\alpha \pi S_{\alpha'}$ with $S_\alpha \cap \pi S_{\alpha'}\pi^{-1} = \{1\}$ in the following sense: If $S_\alpha := H^\alpha$ and if we are given $S_{\alpha'} \leq S_n$, then T^α illustrates permutations ρ (namely the ρ with $\rho S_{\alpha'}\rho^{-1} = V^\alpha$) which satisfy $S_\alpha \cap \rho S_{\alpha'}\rho^{-1} = \{1\}$.

These are the results on symmetric and alternating groups we wished to summarize first.

From now on we shall use the cited results on the symmetric group and its conjugacy classes to define and examine wreath products. We shall return in section 4 to double cosets to construct idempotent elements of the group algebra of S_n which generate minimal left ideals affording irreducible representations of S_n over the field of complex numbers.

2. Wreath products

We define the wreath product of two groups as follows (see e.g. Huppert [1], I § 15):

2.1 Def.: If G is a group, H a permutation group on the set of symbols $\Omega = \{1,\ldots,n\}$, the set

$$\{(f;\pi) \mid f \text{ mapping } \Omega \text{ into } G,\ \pi \in H\}$$

together with the composition law

$$(f;\pi)(f';\pi') := (ff'_{\pi};\pi\pi')$$

is called the <u>wreath product</u> $G \sim H$ <u>of</u> G <u>with</u> H (sometimes <u>G-wreath with</u> H).

Included that to $f:\Omega \to G$ and $\pi \in H$ the mapping $f_{\pi}:\Omega \to G$ is defined by

$$f_{\pi}(\pi(i)) := f(i),\ \forall\ i \in \Omega,$$

and for two mappings $f,f':\Omega \to G$ their product $ff':\Omega \to G$ by

$$ff'(i) := f(i)f'(i),\ \forall\ i \in \Omega\ ,$$

$G \sim H$ is a group as can be seen easily. While checking this one notices, that

2.2 $$(f_{\pi})_{\pi'} = f_{\pi'\pi},\ \forall\ f,\pi,\pi',$$

since products of permutations have to be read here from right to left.

If we denote by $e:\Omega \to G$ the mapping with the values

$$e(i) = 1_G, \ \forall\, i \in \Omega,$$

and if we define to $f:\Omega \to G$ the mapping $f^{-1}:\Omega \to G$ by

$$f^{-1}(i) := f(i)^{-1}, \ \forall\, i \in \Omega,$$

we have for the identity element of $G \sim H$ and for the inverse of $(f;\pi)$:

2.3 $$1_{G\sim H} = (e;1_H) \ , \ (f;\pi)^{-1} = (f^{-1}_{\pi^{-1}};\pi^{-1})$$

$((f_{\pi^{-1}})^{-1}=(f^{-1})_{\pi^{-1}}=: f^{-1}_{\pi^{-1}})$.

The order of this group is

2.4 $$|G\sim H| = |G|^n|H| \ .$$

For the derivation of the representation theory of these groups which we are interested in, the following normal divisor is very important:

2.5 $$G^* := \{(f;1_H) \mid f:\Omega \to G\} = G_1 \times\ldots\times G_n \trianglelefteq G\sim H \ .$$

G^* is called the <u>basis group</u> of $G\sim H$. It is the direct product of n copies G_i of G:

2.6 $$G_i := \{(f;1_H) \mid f(j) = 1_G, \ \forall\, j \neq i\} \simeq G \ .$$

The subgroup

2.7 $$H' := \{(e;\pi) \mid \pi \in H\} \simeq H$$

is the complement of G^* and isomorphic to H, i.e.

2.8 $$G\sim H = G^*H', \ G^* \trianglelefteq G\sim H, \ G^* \cap H' = 1_{G\sim H} = (e;1_H) \ .$$

Let us now look for examples. We shall recognize some subgroups of the symmetric group as permutation representations of certain wreath products. Looking for these examples we may follow the historical development of these ideas.

Such permutation groups arose in the process of constructing a p-Sylow-subgroup of the symmetric group, even before Sylow proved his famous theorems in 1872.

The first description of this construction was probably given by A. Cauchy in the third volume of his "Exercises d'analyse et de physique mathématique" which appeared in 1844 (cf. also the sections 39 and 40 of E. Netto's "Substitutionentheorie und ihre Anwendungen auf die Algebra" (1882) as well as A. Radzig's dissertation "Die Anwendung des Sylow'schen Satzes auf die symmetrische und die alternirende Gruppe" (1895)).

Let us denote by $e_p(m)$ the exponent of the maximal power of a prime number p dividing m. If we assume that $p^s \le n$, $p^{s+1} > n$ and $a_s p^s \le n$ $(0 \le a_s < p)$, but $(a_s+1)p^s > n$, then since

$$n! = 1\cdot 2\cdot \ldots \cdot p^s(p^s+1)\ldots 2p^s \ldots a_s p^s(a_s p^s+1)\ldots(a_s p^s+(n-a_s p^s))$$

we can conclude that

$$e_p(n!) = a_s e_p(p^s!) + e_p((n-a_s p^s)!) \; .$$

If now $a_{s-1}p^{s-1} \le n-a_s p^s$, $(a_{s-1}+1)p^{s-1} > n-a_s p^s$, then in the same way

$$e_p(n!) = a_s e_p(p^s!) + a_{s-1} e_p(p^{s-1}!)$$

and so on.

Hence the problem of constructing a p-Sylow-subgroup of S_n can be reduced to the construction of a p-Sylow-subgroup of symmetric groups S_{p^r} of p-power-degrees p^r.

The maximal power of p in $p^r!$ is

$$p^{e_p(p^r!)} = p^{1+p+p^2+\ldots+p^{r-1}} = p^{(1+p+\ldots+p^{r-2})p}\, p$$

2.9
$$= p^{e_p(p^{r-1}!)p}\, p\,,$$

hence every p-Sylow-subgroup P^r of S_{p^r} has the order

2.10
$$|P^r| = |P^{r-1}|^p\, p\,,$$

if P^{r-1} is a p-Sylow-subgroup of $S_{p^{r-1}}$.

Therefore it would be desirable to construct with the aid of P^{r-1} - which is a permutation group of degree p^{r-1} - a permutation group of degree p^r and of order $|P^{r-1}|^p\, p$. On account of 2.10 the result would be a p-Sylow-subgroup of S_{p^r}.

That this might be done by constructing a faithful permutation representation of $P^{r-1} \sim C_p$ ($C_p := \langle(1\ldots p)\rangle \leq S_p$) is suggested by a comparison of 2.4 and 2.10.

More generally: from two permutation groups G and H of the degrees m and n construct a permutation group of degree mn and of the order $|G|^n|H|$.

We prove first:

2.11 <u>If</u>

$$G \ni g \to \binom{i}{g(i)}$$

<u>is a permutation representation of G on the set of symbols $\Gamma = \{1,\ldots,m\}$ then</u>

$$(f;\pi)(i,j) := (f(\pi(j))(i), \pi(j)),\ \forall\ (i,j) \in \Gamma\times\Omega,$$

<u>yields a permutation representation of $G \sim H$ on $\Gamma \times \Omega$.</u>
<u>This permutation representation of $G \sim H$ is faithful if the permutation representation of G is faithful, and it is transitive if both G (on Γ) and H (on Ω) are transitive.</u>

Proof: a) It is easy to verify, that

$$(f;\pi)((f';\pi')(i,j)) = (ff'_\pi;\pi\pi')(i,j), \ \forall\ f,f',\pi,\pi',i,j\ .$$

And from

$$(f;\pi)(i,j) = (f;\pi)(i',j')$$

we get

$$(f(\pi(j))(i),\pi(j)) = (f(\pi(j'))(i'),\pi(j'))\ .$$

π is a permutation, hence this implies $j = j'$, so that we have

$$f(\pi(j))(i) = f(\pi(j))(i')\ .$$

$f(\pi(j))$ is a permutation, too, hence $i = i'$.

Thus we have obtained a permutation representation of $G \sim H$ as claimed.

b) If $(f;\pi)$ is in the kernel we have

$$(i,j) = (f;\pi)(i,j) = (f(\pi(j))(i),\pi(j)), \ \forall\ i,j\ .$$

If the given permutation representation of G is faithful this implies $f(j) = 1_G$, $\forall\ j$, and hence in this case $f = e$, so that the representation of $G \sim H$ is faithful, too.

c) Finally if G is transitive on Γ, H transitive on Ω, and (i,j) and (i',j') are two symbols out of $\Gamma \times \Omega$, there is a $\pi \in H$ and a $g \in G$ such that $\pi(j) = j'$ and $g(i) = i'$.

If we choose an $f:\Omega \to G$ so that $f(j') = g$, then

$$(f;\pi)(i,j) = (f(\pi(j))(i),\pi(j)) = (f(j')(i),j') = (i',j'),$$

and the permutation representation of $G\sim H$ is transitive as well. This completes the proof.

q.e.d.

The permutation representation of $G\sim H$ given in 2.11 is of degree $|\Gamma\times\Omega| = mn$. Two permutation groups G_1 and G_2 on Ω_1 and Ω_2 are called <u>similar</u>, if there is a bijective mapping ε of Ω_1 onto Ω_2 and an isomorphism φ of G_1 onto G_2 so that

2.12 $$\varepsilon(g(i)) = \varphi(g)(\varepsilon(i)),\ \forall\ i \in \Omega_1, g \in G_1 .$$

We would like to describe a subgroup of S_{mn} similar to the permutation representation of $G\sim H$ given by 2.11.

For such an $\varepsilon:\Gamma\times\Omega \to \Delta = \{1,\ldots,mn\}$ we choose the bijection defined by

2.13 $$\varepsilon(i,j) := (j-1)m + i,\ 1\leq i\leq m,\ 1\leq j\leq n .$$

To describe an isomorphism of the permutation representation of $G\sim H$ into S_{mn} we first divide the set Δ of the symbols on which S_{mn} acts into n pairwise disjoint subsets Δ_i of order m, say

2.14 $$\Delta = \{\underbrace{1,\ldots,m}_{\Delta_1},\underbrace{m+1,\ldots,2m}_{\Delta_2},\ldots,\underbrace{(n-1)m+1,\ldots,nm}_{\Delta_n}\} .$$

Now let σ be a permutation which permutes the Δ_i cyclically, say

2.15 $$\sigma := (1,m+1,\ldots,(n-1)m+1)(2,m+2,\ldots)\ldots(m,2m,\ldots,nm).$$

With the aid of σ we now define an isomorphism φ of the permutation representation of $G \sim H$ into S_{mn} as follows. We define first the images of the values f(i) of f by

2.16 $$\varphi(f(i)) := \sigma^{i-1} f(i) \sigma^{1-i} =: \pi_i ,$$

so that π_i acts on Δ_i, and

2.17 $$\varphi(f;1_H) := \prod_i \pi_i .$$

Now let

2.18 $$\varphi(e;\pi) =: \pi^*$$

be defined as follows:

2.19 $$\pi^*((j-1)m+i) := (\pi(j)-1)m + i, \ 1 \leq i \leq m, \ 1 \leq j \leq n ,$$

so that

2.20 $$\pi^* \Delta_i = \Delta_{\pi(i)} .$$

This means, that the image of $(e;\pi)$ under φ is the permutation of the Δ_i corresponding to the permutation π of the symbols i of $\Omega = \{1,\ldots,n\}$. As can easily be seen, this mapping φ defined by

2.21 $$\varphi(f;\pi) := \pi_1 \ldots \pi_n \pi^*$$

is an isomorphism of the permutation representation of $G \sim H$ described in 2.11 into S_{mn}. The image of the permutation representation is obviously

2.22 $$(G_1 \times \ldots \times G_n)H' \leq S_{mn} ,$$

with G_1 equal to the permutation representation of G and

2.23 $$G_i = \sigma^{i-1} G_1 \sigma^{1-i} .$$

H' is the permutation group acting on the subsets Δ_i of Δ as H acts on the symbols i of Ω; H' is a subgroup of the subgroup

$S_n' \leq S_{mn}$ which consists of the n! permutations of the Δ_i.

To prove the similarity we have to verify, that 2.12 is valid. Starting with the left hand side of 2.12 we obtain

$$\varepsilon((f;\pi)(i,j)) = \varepsilon(f(\pi(j))(i),\pi(j)) = (\pi(j)-1)m + f(\pi(j))(i) \ .$$

And on the right hand side we have

$$\varphi(f;\pi)(\varepsilon(i,j)) = \pi_1 \ldots \pi_n \pi^*((j-1)m + i)$$
$$= \pi_1 \ldots \pi_n((\pi(j)-1)m + i) = (\pi(j)-1)m+f(\pi(j))(i) \ .$$

Thus we have proved the following:

2.24 The mappings ε and φ defined by 2.13–2.21 map the permutation representation of G∿H described by 2.11 onto a similar subgroup of S_{mn}.

It is interesting to see, that these groups are the permutation groups used by Cauchy and Netto to construct Sylow-subgroups of the symmetric group and which arose at the beginning of the development of the concept of the wreath product of groups. A sketch of this construction reads as follows:

2.25 A faithful permutation representation of degree mn of the wreath product of two permutation groups G (on $\Gamma=\{1,\ldots,m\}$) and H (on $\Omega = \{1,\ldots,n\}$) can be obtained as follows: Divide the set $\Delta = \{1,\ldots,mn\}$ on which S_{mn} acts into disjoint subsets $\Delta_1=\Gamma$, Δ_2, ..., Δ_n of order m. Form the direct product of $G = G_1$ (on $\Delta_1=\Gamma$) with the subgroups G_i (on the subsets Δ_i) which correspond to G_i ($2 \leq i \leq n$) and multiply $G_1 \times \ldots \times G_n$ with the subgroup H' corresponding to H and acting on the subsets Δ_i as H acts on the symbols i of Ω. The result

$$(G_1 \times \ldots \times G_n)H'$$

is the desired faithful permutation representation of $G \sim H$ and it is transitive if both G (on Γ) and H (on Ω) are transitive.

A simple example is the representation of $S_2 \sim S_2 = C_2 \sim C_2$ (if we denote by C_n the cyclic group of order n):

$$C_2 \sim C_2 = \{(1,1;1),(1,(12);1),((12),1;1),((12),(12);1),(1,1;(12)),$$
$$(1,(12);(12)),((12),1;(12)),((12),(12);(12))\}$$

($(f;\pi)$ written in the explicit form $(f(1),f(2);\pi)$) is similar to

$$(\{1,(12)\} \times \{1,(34)\})\{1,(13)(24)\}$$
$$= \{1,(12),(34),(12)(34),(13)(24),(14)(23),(1324),(1423)\}.$$

This image of $C_2 \sim C_2$ is a 2-Sylow-subgroup of S_4.

Coming back to the p-Sylow-subgroups we get from these considerations:

2.26 If P^{r-1} is a p-Sylow-subgroup of $S_{p^{r-1}}$ on $\Gamma = \{1,\ldots,p^{r-1}\}$, and if $C_p = \langle(1\ldots p)\rangle \leq S_p$, then the subgroup of S_{p^r} similar to $P^{r-1} \sim C_p$ as described in 2.25 is a p-Sylow-subgroup.

Using this result we can construct a p-Sylow-subgroup of S_n.

If

$$n = \sum_{i=0}^{t} a_i p^i \ , \ 0 \leq a_i < p,$$

then as can be seen from considerations above the exponent of the maximal power of p dividing n! is

2.27
$$\begin{aligned} e_p(n!) &= a_1 + a_2(1+p) + \ldots + a_t(1+p+\ldots+p^{t-1}) \\ &= a_1 e_p(p!) + a_2 e_p(p^2!) + \ldots + a_t e_p(p^t!) \ . \end{aligned}$$

Let us now divide the set $\Omega = \{1,\ldots,n\}$ on which S_n acts into pairwise disjoint subsets as follows. We divide Ω into the subset $\Omega_o := \{1,\ldots,a_o\}$ of order a_o, into the a_1 subsets $\Omega_{11} := \{a_o+1,\ldots,a_o+p\}$, ..., $\Omega_{1a_1} := \{a_o+(a_1-1)p+1,\ldots,a_o+a_1p\}$ of order p, ..., a_t subsets $\Omega_{t1},\ldots,\Omega_{ta_t}$ of order p^t. On each of these subsets Ω_{ij} form a p-Sylow-subgroup $P^{i-1}\wr C_p$ of S_{p^i} (regarded as subgroup of S_n). And now take their direct product

$$\mathop{\times}_{i=1}^{t} (\mathop{\times}^{a_i} (P^{i-1}\wr C_p)) \leq S_n$$

$(\mathop{\times}^{a_i} (P^{i-1}\wr C_p) := P^{i-1}\wr C_p \times\ldots\times P^{i-1}\wr C_p$ (a_i factors)).

The resulting subgroup is of order $p^{e_p(n!)}$ and hence a p-Sylow-subgroup of S_n. Thus we have derived the well-known construction of p-Sylow-subgroups first given by Cauchy (for this and other results concerning p-Sylow-subgroups of symmetric groups cf. also the papers of L. Kaloujnine and the paper of Weir (see the references)):

<u>2.28 If $n = \Sigma\, a_i p^i$ $(0 \leq a_i < p)$, then every p-Sylow-subgroup of S_n is of the form $\underset{i}{\times}\, (\overset{a_i}{\times} (P^{i-1} \wr C_p))$.</u>

This as well as 2.26 gives a recursion formula for the construction of a p-Sylow-subgroup. Because of the associativity of the wreath product multiplication this can be written explicitly:

<u>2.29 Let G, H and I be permutation groups on Γ, Ω and Δ. Then $(G \wr H) \wr I$ and $G \wr (H \wr I)$ are corresponding permutation groups, if we identify $(\Gamma \times \Omega) \times \Delta$ and $\Gamma \times (\Omega \times \Delta)$ according to</u>

$$((i,j),k) = (i,(j,k)) \ .$$

Proof: Let $(f;\pi)$ be an element of $(G \wr H) \wr I$ so that $f(i) = (f_i;\pi_i) \in G \wr H$, $i \in \Delta$. And let $(f^*;\pi^*)$ be an element of $G \wr (H \wr I)$ with $\pi^* = (f';\pi) \in H \wr I$ so that $f'(i) = \pi_i$ and $f^*(i,j) = f_j(i)$. Then on the one hand

$$\begin{aligned} (f;\pi)((i,j),k) &= ((f_{\pi(k)};\pi_{\pi(k)})(i,j),\pi(k)) \\ &= ((f_{\pi(k)}(\pi_{\pi(k)}(j))(i),\pi_{\pi(k)}(j)),\pi(k)) \ . \end{aligned}$$

And on the other hand

$$\begin{aligned} (f^*;\pi^*)(i,(j,k)) &= (f^*((f';\pi)(j,k))(i),(f';\pi)(j,k)) \\ &= (f^*(f'(\pi(k))(j),\pi(k))(i),(f'(\pi(k))(j),\pi(k))) \\ &= (f_{\pi(k)}(\pi_{\pi(k)}(j))(i),(\pi_{\pi(k)}(j),\pi(k)) \ . \end{aligned}$$

Thus $(f;\pi)$ acts as $(f^*;\pi^*)$ if we identify the symbols as described.

q.e.d.

A corollary is

2.30 If $C_p := \langle(1\ldots p)\rangle \leq S_p$, then every p-Sylow-subgroup of S_{p^r} is similar to the wreath product

$$\overset{r}{\sim} C_p := C_p \sim \ldots \sim C_p \text{ (r factors)} .$$

Every p-Sylow-subgroup of S_n ($n = \Sigma\, a_i p^i$, $0 \leq a_i < p$) is similar to

$$\underset{i}{\times} \left(\overset{a_i}{\times} \left(\overset{r}{\sim} C_p \right) \right) .$$

As we have seen, Sylow-subgroups of symmetric groups are direct products of wreath products. In the introduction we claimed, that this is also valid for the centralizers of elements in S_n.
To show this, we start with a special case, namely the centralizer of the permutation

$$\pi = \pi_1 \ldots \pi_n := (1\ldots m)(m+1,\ldots,2m)\ldots(\ldots nm) \in S_{mn} .$$

The centralizer $C_{S_{mn}}(\pi)$ of this permutation is

2.31 $$C_{S_{mn}}(\pi) = (\langle\pi_1\rangle \ldots \langle\pi_n\rangle)S_n' .$$

That means $C_{S_{mn}}(\pi)$ is the product of the subgroup generated by the cyclic factors π_i of π with the subgroup S_n' of the n! permutations of S_{mn} which permute the sets of symbols in the cyclic factors of π as they stand.
Following the considerations preceeding 2.24 and 2.25 we see, that this subgroup is a permutation representation of $C_m \sim S_n$.
And applying this to the subsets of cyclic factors of the same

length in a general permutation we have:

2.32 If $\pi \in S_n$ is of type $T\pi = (a_1,\ldots,a_n)$, then the centralizer of π in S_n is a faithful permutation representation of

$$\underset{i}{\times} (C_i \sim S_{a_i})$$

$(C_i \sim S_o := \{1\},\ C_i := \langle(1\ldots i)\rangle \leq S_i)$.

(Remark: From this we can derive at once the result 1.23 about the splitting of S_n-classes into A_n-classes.) Analogously, we have the following result which will be of use later on:

2.33 If S_{mn} is the symmetric group on $\Omega = \{1,\ldots,mn\}$, then the normalizer $N_{S_{mn}}(\overset{n}{\times} S_m)$ of the subgroup

$$\overset{n}{\times} S_m := S_m \times\ldots\times S_m \text{ (n factors)}$$

(i-th factor S_m on $\Omega_i := \{(i-1)m+1,\ldots,im\}$) is the faithful permutation representation of $S_m \sim S_n$ described in 2.24 and 2.25.

These three examples, the p-Sylow-subgroups, the centralizers of elements and the normalizers of subgroups of the form $\overset{n}{\times} S_m$ in symmetric groups show, how useful this concept of the wreath product is. Moreover they give a hint as to how the representation theory of wreath products may be applied to the representation theory of the symmetric group.

The centralizers of p-elements as well as the p-Sylow-subgroups of the centralizers of p-regular elements play an important role in the p-modular representation theory of finite groups. On the other hand it is known, that the so-called symmetrized outer products of two irreducible representations of S_m and S_n are representations of S_{mn} induced by certain irreducible representations of $N_{S_{mn}}(\overset{n}{\times} S_m)$.

The other way round we may ask how the representation theory of the symmetric group can be applied to derive the representation theory of certain wreath products.

About 1930 A. Young applied his methods to the so-called hyperoctahedral groups. In our notation these are groups of the form $S_2 \sim S_n$, and they arise by representing the elements of S_n by permutation matrices and allowing not only +1 as nonvanishing entries but also -1 (cf. Young [1]).

Following a suggestion of I. Schur, W. Specht then considered in his dissertation such groups where not only ± 1 are allowed as nonvanishing entries but even the elements of a group G if the matrix multiplication is defined appropriately (Specht [1]). These groups are obviously of the form $G \sim S_n$. In a following paper (Specht [2]) he considered the general case $G \sim H$, H a finite permutation group. He showed that especially the ordinary representation theory of wreath products of the form $G \sim S_n$ can be largely derived with the aid of the representation theory of S_n.

Nevertheless the application of the theory of the representations of wreath products to the representation theory of the symmetric group is not a vicious circle. For those wreath products $G\sim S_m$ whose representations we shall apply to S_n satisfy $m<n$, so that on the contrary this application provides an interesting recursion process.

In order to derive the representation theory of wreath products of the form $G\sim S_n$ it is necessary to examine such groups more closely. This we shall do in the following section.

3. Wreaths with symmetric groups

We shall now consider wreath products of the form $G\sim S_n$, - wreaths with symmetric groups. For the time being let G denote a finite group.

Since $G\sim S_n$ may arise as described at the end of the last section by inserting elements of G for the nonvanishing entries of permutation matrices representing the elements of S_n, the group $G\sim S_n$ is sometimes called the complete monomial group (of degree n) of G. O. Ore called it symmetry (of degree n) of G (Ore [1]).

Wreaths $C_m\sim S_n$ of cyclic groups with symmetric groups have been called generalized symmetric groups by M. Osima (Osima [1]), and analogously the groups $C_m\sim A_n$ were called generalized alternating groups (Puttaswamaiah [1]). As has been mentioned above, the special case $C_2\sim S_n$ is called a hyperoctahedral group.

Let us first consider the conjugacy classes of $G\sim S_n$, which have been characterized by W. Specht (Specht [1]).

Since $\{1\}\sim S_n \simeq S_n$ this characterization provides a generalization of the characterization of the conjugacy classes of symmetric groups by their cycle decomposition. We choose the notation so that we shall obtain a generalization of the characterization by the type of the permutation.

Let $(f;\pi)$ be an element of $G\wr S_n$. The cycle-notation of π is uniquely determined if we agree to begin each cyclic factor with the least symbol included in it. With this convention we can associate with each cyclic factor $(j\ \pi(j)\ldots\pi^r(j))$ of π with respect to $f:\Omega\to G$ the uniquely determined element

3.1 $$ff_\pi\ldots f_{\pi^r}(j) = f(j)f(\pi^{-1}(j))\ldots f(\pi^{-r}(j)) \in G\ .$$

We call this element <u>the cycleproduct associated with</u> $(j\ldots\pi^r(j))$ <u>with respect to</u> f.

This done we can give the following definition which generalizes the concept of the type of a permutation to the concept of the type of $(f;\pi)$. As we shall see soon afterwards, this type characterizes a conjugacy class of $G\wr S_n$.

<u>3.2 Def.:</u> Let the permutation $\pi\in S_n$ be of type $T\pi=(a_1,\ldots,a_n)$, and let f be a mapping of Ω into G such that $(f;\pi)\in G\wr S_n$.

There are a_k cycleproducts associated with the a_k cycles of length k of π with respect to f as described in 3.1.

Let $C^1,\ldots,C^s$ be an ordering of the conjugacy classes of the finite group G.

If now exactly a_{ik} of these a_k cycleproducts belong to C^i we call the $(s\times n)$-matrix

$$T(f;\pi) := (a_{ik}) \qquad \begin{array}{l} i \text{ row index, } 1\le i\le s, \\ k \text{ column index, } 1\le k\le n, \end{array}$$

the <u>type of</u> $(f;\pi)$.

In the special case s=1, i.e. in case $\{1\}\sim S_n = S_n$, we have the row $(a_1,\ldots a_n)$ and hence 3.2 is a generalization of the definition 1.16 of the type of a permutation.

We would like to show that as in 1.14/1.16, two elements of $G\sim S_n$ belong to the same conjugacy class if and only if they are of the same type.

Before we can prove this we need some preliminary considerations. At first we notice, that the entries a_{ik} of the type (a_{ik}) of an element $(f;\pi) \in G\sim S_n$, with π of type $(a_1,\ldots,a_n)$, satisfy the following equations:

3.3 $$0 \leq a_{ik} \in \mathbb{Z}, \quad \sum_i a_{ik} = a_k, \quad \sum_{i,k} ka_{ik} = n .$$

To each (sxn)-matrix (a_{ik}) whose entries satisfy 3.3, there are elements in $G\sim S_n$, which are of this type, since f ranges over all the mappings of Ω into G.

Since we are merely interested in the type, we need only determine a cycleproduct up to conjugation in G. Therefore we show first, that the convention that the symbol j in 3.1 is the least symbol of the considered cycle is unnecessary, i.e. that

<u>3.4</u> $$f\ldots f_{\pi^r}(j) \sim f\ldots f_{\pi^r}(\pi^s(j)), \; \forall\, s \in \mathbb{N} .$$

Proof: We can assume $0 \leq s \leq r$. Then

$$f\ldots f_{\pi^r}(\pi^s(j)) = f(\pi^s(j))\ldots f(\pi(j))f(j)\ldots f(\pi^{s+1}(j)). \quad (1)$$

If

$$a := f(\pi^s(j))\ldots f(\pi(j)), \quad b := f(j)f(\pi^{-1}(j))\ldots f(\pi^{s+1}(j)),$$

then ab is the right hand side of the equation (1), and ba is the cycleproduct to $(j \dots \pi^r(j))$ with respect to f (cf. 3.1). But $ab \sim ba$: $a^{-1}aba = ba$.

q.e.d.

Another useful remark is

3.5
$$\begin{aligned} T(f;\pi) &= T(e;\pi')(f;\pi)(e;\pi')^{-1} \\ &= T(f';1)(f;\pi)(f';1)^{-1} \end{aligned} \qquad , \forall\, f,f',\pi,\pi'.$$

Proof: We have

$$(e;\pi')(f;\pi)(e;\pi')^{-1} = (f_{\pi'};\pi'\pi\pi'^{-1}),$$

$$(f';1)(f;\pi)(f';1)^{-1} = (f'ff_{\pi}'^{-1};\pi).$$

Hence it suffices to prove, that the right hand sides of these two equations are elements of type $T(f;\pi)$.

To decide this, it suffices to show, that the cycleproduct g to the cyclic factor $(j \dots \pi^r(j))$ of π with respect to f is conjugate to the following two cycleproducts: the cycleproduct g' belonging to the cyclic factor $(\pi'(j) \dots \pi'\pi^r(j))$ of $\pi'\pi\pi'^{-1}$ with respect to $f_{\pi'}$ and the cycleproduct g" associated with $(j \dots \pi^r(j))$ with respect to $f'ff_{\pi}'^{-1}$.

Using 3.4 we obtain

$$\begin{aligned} g' &\sim f_{\pi'}(f_{\pi'})_{\pi'\pi\pi'^{-1}} \dots (f_{\pi'})_{\pi'\pi^r\pi'^{-1}}(\pi'(j)) \\ &= f_{\pi'}f_{\pi'\pi} \dots f_{\pi'\pi^r}(\pi'(j)) = g . \end{aligned}$$

And for g" we have

$$g'' = f'ff_{\pi}'^{-1}f_{\pi}'f_{\pi}f_{\pi^2}'^{-1} \dots f_{\pi^r}'f_{\pi^r}f_{\pi^{r+1}}'^{-1}(j)$$

$= f'(j)gf'(j)^{-1} \sim g$.

q.e.d.

A last lemma before we prove, that the type characterizes a conjugacy class:

3.6 <u>If $T(f;\pi) = T(f';\pi')$, then there is a $\pi'' \in S_n$ satisfying $\pi = \pi''\pi'\pi''^{-1}$ and with the property, that for each cyclic factor of π the two cycleproducts with respect to f and with respect to $f'_{\pi''}$ are conjugates.</u>

Proof: If $T(f;\pi) = T(f';\pi')$ we obtain from 3.3: $\pi \sim \pi'$. Hence there is a $\tilde{\pi} \in S_n$, which satisfies

$$\pi = \tilde{\pi}\pi'\tilde{\pi}^{-1}.$$

The set of all these $\tilde{\pi}$ forms a right coset of the centralizer of π. Since

$$(e;\tilde{\pi})(f';\pi')(e;\tilde{\pi})^{-1} = (f'_{\tilde{\pi}};\pi) ,$$

3.5 implies

$$T(f;\pi) = T(f';\pi') = T(f'_{\tilde{\pi}};\pi) . \tag{1}$$

The cycleproduct to the factor $(j\ldots\pi^r(j))$ of π with respect to the mapping f is

$$f\ldots f_{\pi^r}(j) ,$$

and with respect to $f'_{\tilde{\pi}}$:

$$f'_{\tilde{\pi}}\ldots f'_{\pi^r\tilde{\pi}}(j) .$$

Since (1) is valid, there is a π^* from the centralizer of π, which yields, if it is applied to $f'_{\tilde{\pi}}$:

$$f'_{\pi^*\tilde{\pi}} \ldots f'_{\pi^r\pi^*\tilde{\pi}}(j) \sim f \ldots f_{\pi^r}(j) \ .$$

Thus $\pi'' := \pi^*\tilde{\pi}$ fulfils the statement.

q.e.d.

Now we are ready to characterize the conjugacy classes (Specht [1]):

<u>3.7</u> $$(f;\pi) \sim (f';\pi') \Leftrightarrow T(f;\pi) = T(f';\pi')$$

Proof:

a) If $(f;\pi) \sim (f';\pi')$, then there are f'' and π'' so that

$$(f'';\pi'')(f;\pi)(f'';\pi'')^{-1} = (f'';1)(e;\pi'')(f;\pi)(e;\pi'')^{-1}(f'';1)^{-1}$$
$$= (f';\pi') \ ,$$

and using 3.5 we obtain $T(f;\pi) = T(f';\pi')$ as claimed.

b) If the other way round $T(f;\pi) = T(f';\pi')$, then by 3.6 there is a $\pi'' \in S_n$ satisfying $\pi = \pi''\pi'\pi''^{-1}$. Now

$$(e;\pi'')(f';\pi')(e;\pi'')^{-1} = (f'_{\pi''};\pi''\pi'\pi''^{-1}) = (f'_{\pi''};\pi) \ .$$

Because of 3.5, this implies $T(f';\pi') = T(f'_{\pi''};\pi)$. And since these two elements are conjugates, it suffices to show, that

$$(f'_{\pi''};\pi) \sim (f;\pi) \ . \qquad (2)$$

We assume now that π'' has been chosen in such a way, that for each cyclic factor $(j \ldots \pi^r(j))$ of π

$$f \ldots f_{\pi^r}(j) \sim f'_{\pi''} \ldots f'_{\pi^r\pi''}(j)$$

(cf. 3.6).

Let g_j be an element of G such that

$$f \ldots f_{\pi^r}(j) = g_j(f'_{\pi''} \ldots f'_{\pi^r \pi''}(j))g_j^{-1} .$$

Having chosen such a g_j and starting with $u = 0$ we obtain from the equations

$$f(\pi^{-u}(j)) = g_{\pi^{-u}(j)} f'_{\pi''}(\pi^{-u}(j)) g^{-1}_{\pi^{-u-1}(j)} \tag{2}$$

elements $g_{\pi^{-u-1}(j)}$, which are uniquely determined ($1 \leq u \leq r$), if $g_j = g_{\pi^{-0}(j)}$ is fixed.

Having done this for every cyclic factor of π we define the mapping $f^*: \Omega \to G$ by

$$f^*(i) = g_i, \ \forall \ i \in \Omega .$$

And this mapping satisfies

$$(f^*;1)(f'_{\pi''};\pi)(f^*;1)^{-1} = (f^* f'_{\pi''} f^{*-1}_{\pi};\pi) = (f;\pi)$$

(the last equation follows from (2)), which proves (1).

q.e.d.

Hence $G \sim S_n$ contains exactly as many conjugacy classes as there are types, i.e. (sxn)-matrices (a_{ik}) whose entries satisfy 3.3. For the number of types or conjugacy classes we prove now (Specht [1]):

<u>3.8</u> <u>If $p(m)$ is the number of partitions of m for $m \in \mathbb{N}$ and $p(0) := 1$, the number of conjugacy classes of $G \sim S_n$ is</u>

$$\sum_{(n)} p(n_1) \ldots p(n_s) ,$$

<u>if the sum is taken over all the s-tupels $(n) = (n_1, \ldots, n_s)$ (s=number of conjugacy classes of G) with $\Sigma\, n_i = n$, $0 \leq n_i \in \mathbb{Z}$.</u>

Proof: For a type (a_{ik}) we define $n_i := \sum_k ka_{ik}$. Then $(n_1,\ldots,n_s)$ is such an s-tupel, and all the s-tupels occur in this way. n_i is the sum of the elements of the i-th row of (a_{ik}) weighted with their column number. Therefore if the other rows are fixed, there are exactly $p(n_i)$ possibilities for the i-th row to be the row of a type. And this proves the assertion.

q.e.d.

If we wish to derive the order of such a conjugacy class, say of the conjugacy class of $G\wr S_n$ which is characterized by the type (a_{ik}), we set

$$a_k := \sum_i a_{ik}\ ,\ 1\leq k\leq n\ .$$

Then there are

$$n!/\prod_k k^{a_k} a_k!$$

elements of type $(a_1,\ldots,a_n)$ in S_n. We choose one of them, say π. The a_k cyclic factors of π which are of length k can be distributed into the s conjugacy classes of G in

$$\binom{a_k}{a_{1k}}\binom{a_k-a_{1k}}{a_{2k}}\cdots\binom{a_k-a_{1k}-a_{2k}-\cdots-a_{s-1,k}}{a_{sk}} = \frac{a_k!}{a_{1k}!\cdots a_{sk}!}$$

ways which are in accordance with the considered type (a_{ik}). Let $f:\Omega\to G$ be a mapping which yields such a distribution of the cycleproducts. It remains to show, what freedom of choice is left for choosing the values of f.

To assure that

$$f \ldots f_{\pi^{k-1}}(j) \in C^i$$

we may choose the values $f(j)$, $f(\pi^{-1}(j))$, ..., $f(\pi^{-k+2}(j))$ at will and can choose an $f(\pi^{-k+1}(j)) \in G$ so that the complete product is an element of $C^i \subseteq G$. Hence there exist

$$\frac{a_k!}{a_{1k}!\ldots a_{sk}!} = (|G|^{k-1}|C^i|)^{a_k}$$

mappings $f:\Omega \to G$ which distribute the a_k k-cycles of π as the considered type (a_{ik}) prescribes.

We have to multiply this number of mappings with the number of elements of type $(a_1,\ldots,a_n)$ and take the product over all i and k to obtain the order of the conjugacy class of $G\sim S_n$ which is characterized by the considered type (a_{ik}). Thus we have (Specht [1]):

3.9 <u>The conjugacy class of $G\sim S_n$ consisting of the elements of the elements of type (a_{ik}) has the order</u>

$$|G\sim S_n| / \prod_{i,k} a_{ik}!(k|G|/|C^i|)^{a_{ik}} .$$

Therefore the order of the centralizer of an element of type (a_{ik}) is

$$\prod_{i,k} a_{ik}!(k|G|/|C^i|)^{a_{ik}}.$$

Before we can describe this centralizer in detail we have to consider the orders of the elements.

At first we notice, that for $u \in \mathbb{N}$:

3.10 $$(f;\pi)^u = (ff_\pi \ldots f_{\pi^{u-1}};\pi^u) ,$$

what implies, that the order of $(f;\pi)$ is a multiple of the order of π. The order of π is the least common multiple of the lengths of the cyclic factors of π (cf. 1.11).

On the other hand we let $u \in \mathbb{N}$ be a multiple of the length of every cyclic factor of π. If $i \in \Omega$, then there is such a $\nu \geq 0$ that $i = \pi^\nu(j)$ and j is the least symbol of the cyclic factor of π which includes this symbol i. Let g be the cycleproduct associated with this cycle and with respect to f. Then we have

3.11 $$\begin{aligned} f \ldots f_{\pi^{u-1}}(i) &= f(\pi^\nu(j))f(\pi^{\nu-1}(j)) \ldots f(\pi^{\nu-u+1}(j)) \\ &= f(\pi^\nu(j)) \ldots f(\pi(j))g^w f(\pi(j))^{-1} \ldots f(\pi^\nu(j))^{-1} \sim g^w , \end{aligned}$$

if $u = wr$ for the length r of the cycle including i.

The order of $(f;\pi)$ is the minimal $u \in \mathbb{N}$ so that $\pi^u = 1_{S_n}$ and $f \ldots f_{\pi^{u-1}}(i) = 1_G$, $\forall\, i \in \Omega$, i.e. $f \ldots f_{\pi^{u-1}} = e$. Hence from 3.11 we obtain

3.12 <u>The order of $(f;\pi)$ (of type (a_{ik})) is the least common multiple of the products $k\omega_i$ of the lengths k of the cyclic factors of π with the orders ω_i of the corresponding cycleproducts with respect to f:</u>

$$|\langle(f;\pi)\rangle| = \underset{i,k:a_{ik}>0}{\text{l.c.m.}} \{k\omega_i\} .$$

Hence $(f;\pi)$ is p-regular (i.e. $p \nmid |\langle(f;\pi)\rangle|$) if p does not divide any one of these products $k\omega_i$. We have the following corollary:

<u>3.13 The p-regular classes of $G\sim S_n$ are exactly the conjugacy classes belonging to types (a_{ik}) where nonvanishing entries occur only in rows to p-regular classes C^i of G and columns with p-regular numbers k.</u>

<u>The p-classes correspond to those types (a_{ik}) where only in columns with p-power-numbers k and only in rows belonging to p-classes C^i of G do vanishing entries occur.</u>

<u>The remaining types belong to p-singular classes.</u>

3.12 and 3.13 are generalizations of 1.11 and 1.12: for G = {1}, i.e. for one-rowed types we obtain 1.11 and 1.12 at once.
As is well-known, the number of representations of a finite group over the field of complex numbers which have real character is equal to the number of ambivalent conjugacy classes (i.e. classes containing the inverse of each of its elements). As we have seen in section 1, each conjugacy class of a symmetric group is ambivalent, while only the alternating groups A_1, A_2, A_5, A_6, A_{10} and A_{14} are ambivalent (cf. 1.15/1.25). This implies that all the ordinary characters of S_n and all the ordinary characters of these six alternating groups are real. We show, that this is the case for $G\sim S_n$, too, if it is valid for G (Kerber [7]):

<u>3.14 If G is ambivalent, then $G\sim S_n$ is ambivalent.</u>

Proof: There is obviously a 1-1-correspondence between the cycles

$(j \ldots \pi^r(j))$ of π and $(j \ldots \pi^{-r}(j))$ of π^{-1}.

Let g be the cycle product to $(j \ldots \pi^r(j))$ with respect to f.

Then the cycleproduct to $(j \ldots \pi^{-r}(j))$ with respect to $f^{-1}_{\pi^{-1}}$ is

$$f^{-1}_{\pi^{-1}} f^{-1}_{\pi^{-2}} \ldots f^{-1}(j) = g^{-1}$$

(recall that $(f;\pi)^{-1} = (f^{-1}_{\pi^{-1}};\pi^{-1})$).

If now G is an ambivalent group, then $g \sim g^{-1}$, $\forall\, g \in G$, what implies, that in this case

$$T(f;\pi) = T(f;\pi)^{-1}.$$

And this proves the assertion since two elements of the same type are conjugates (cf. 3.7).

q.e.d.

A special case is

3.15 $\forall\, m,n$: $S_m \sim S_n$ is ambivalent.

3.14/3.15 generalize a result of Berggren (Berggren [1]). He proved the ambivalency of $G \sim S_2$ assuming that G is ambivalent. Using this and the associativity of the wreath product multiplication he showed, that the 2-Sylow-subgroups of symmetric groups are ambivalent. And this implies, that every 2-group can be embedded in an ambivalent 2-group.

Cyclic groups C_p of order p are not ambivalent in general, therefore that not every p-Sylow-subgroup of S_n is ambivalent is implied by the following:

3.16 <u>The ambivalency of $G\sim H$ implies the ambivalency of G and the ambivalency of H.</u>

Proof: If $G\sim H$ is ambivalent, then for every $(f;\pi) \in G\sim H$ there is an $(f';\pi')$ so that

$$(f';\pi')(f;\pi)(f';\pi')^{-1} = (f'f_{\pi'}f'^{-1}_{\pi'\pi\pi'^{-1}};\pi'\pi\pi'^{-1}) = (f^{-1}_{\pi^{-1}};\pi^{-1}).(1)$$

This implies, that every $\pi \in H$ is conjugate to its inverse, i.e. H is ambivalent.

And if we choose a constant $f:\Omega \to G$, say

$$f(i) = g, \ \forall\ i \in G,$$

then for $\pi = 1$ we obtain from (1), that there is an f' so that $f'ff'^{-1} = f^{-1}$ what implies $g \sim g^{-1}$. G has to be ambivalent as well.

q.e.d.

Using 1.25 we have the corollary:

3.17 <u>$A_i\sim S_m$ and $S_m\sim A_i$ with $i \in \{1,2,5,6,10,14\}$ are the only ambivalent wreath products of alternating with symmetric groups.</u>

Before we conclude this section with an example let us describe the centralizer of an element $(f;\pi) \in G\sim S_n$.

As in the case of the description of the centralizer of a permutation in S_n (cf. section 2, 2.31/2.32) we start with the consideration of a special case:

$$(f;\pi) \in G\sim S_n \ , \quad T\pi = (0,\ldots,0,1) \ ,$$

i.e. π is a cycle of length n.

In this case the type (a_{ik}) of $(f;\pi)$ has exactly one nonvanishing entry, a 1 in the last column.

Since the centralizers of conjugates are conjugate subgroups, we can assume, that

$$(f;1) \in G_1, \text{ i.e. } f(i) = 1_G, \forall\, i \neq 1,\ f(1) \in C^i,$$

if a_{in} is this only nonvanishing entry of (a_{ik}).

A subgroup of the centralizer of $(f;\pi)$ is the cyclic subgroup $\langle(f;\pi)\rangle \leq G\sim S_n$, generated by $(f;\pi)$ itself. But $(f;\pi)$ commutes also with the elements $(f';1) \in G^*$ whose mappings f' are constant on Ω and so that their value is an element of the centralizer of $f(1)$ in G:

$$f'\colon f'(i) = g \in C_G(f(1)),\ \forall\, i \in \Omega .$$

This follows from

$$(f';1)(f;\pi)(f';1)^{-1} = (f'ff_\pi'^{-1};\pi) = (f'ff'^{-1};\pi) = (f;\pi) .$$

The subgroup of these $(f';1)$ is the diagonal of the basis group of the subgroup $C_G(f(1))\sim S_n \leq G\sim S_n$:

$$\{(f';1)\} = \operatorname{diag}(C_G(f(1))^*) \leq C_G(f(1))^* \leq G^* .$$

Let us multiply these two subgroups of the centralizer and we obtain

3.18 $$\operatorname{diag}(C_G(f(1))^*)\langle(f;\pi)\rangle.$$

We would like to show, that this is a subgroup and of the same order as $C_{G\sim S_n}(f;\pi)$ and hence equal to this centralizer.

If $(f';1), (f'';1) \in \operatorname{diag}(C_G(f(1))^*)$, we have for $r,s \in \mathbb{Z}$:

$$(f';1)(f;\pi)^r((f'';1)(f;\pi)^s)^{-1} = (f';1)(f;\pi)^{r-s}(f''^{-1};1)$$
$$= (f'f''^{-1};1)(f;\pi)^{r-s} .$$

Hence this subset 3.18 is a subgroup.

If we now assume, that

$$(f';1)(f;\pi)^r = (f'';1)(f;\pi)^s$$

we have

$$(f''^{-1}f';1) = (f;\pi)^{s-r} .$$

This is fulfilled only if $s-r \equiv 0$ (n).

On the other hand we have

$$(f;\pi)^n = (ff_\pi \ldots f_{\pi^{n-1}};\pi^n) = (f(1),\ldots,f(1);1) \in \mathrm{diag}(C_G(f(1))^*)$$

and hence if $r \equiv t$ (n) there is an $(f';1) \in \mathrm{diag}(C_G(f(1))^*)$ satisfying

$$(f;\pi)^r = (f';1)(f;\pi)^t.$$

Thus the order of this subgroup of $C_{G\sim S_n}(f;\pi)$ is

$$|\mathrm{diag}(C_G(f(1))^*)\langle(f;\pi)\rangle| = (|G|/|C^i|)n,$$

and this is the order of the centralizer of $(f;\pi)$, as can be seen from 3.9.

Therefore we have proved the following first step towards a construction of the centralizer of a general element of $G\sim S_n$ (Ore [1]):

<u>3.19</u> <u>If π is an n-cycle and $(f;1) \in G_1$, then the centralizer of $(f;\pi)$ in $G\sim S_n$ is the product of the diagonal subgroup of the basis group $C_G(f(1))^*$ of $C_G(f(1))\sim S_n \leq G\sim S_n$ and the</u>

<u>cyclic subgroup generated by $(f;\pi)$, i.e.</u>

$$C_{G\sim S_n}(f;\pi) = \mathrm{diag}(C_G(f(1))^*)\langle(f;\pi)\rangle$$

<u>in this special case.</u>

Using this we would like to describe the centralizer of a general element $(f;\pi) \in G\sim S_n$.

Let us denote by π^j_{ik} $(1\leq j\leq a_{ik})$ the a_{ik} cyclic factors of π whose cycleproducts with respect to f are elements of $C^i \subseteq G$ (if $a_{ik}>0$).

Let π^j_{ik} be the cycle

3.20 $$\pi^j_{ik} = (r^j_{ik}\ \pi(r^j_{ik})\ldots\pi^{k-1}(r^j_{ik}))$$

so that r^j_{ik} is the least symbol of this cycle.

We notice that

3.21 $$\pi = \prod_{i,j,k} \pi^j_{ik} .$$

Without loss of generality we can assume, that f is of the following form:

3.22 $$f(r^j_{ik}) \in C^i,\ f(\pi^{-1}(r^j_{ik})) = \ldots = f(\pi^{-k+1}(r^j_{ik})) = 1_G.$$

If we now define a mapping $f^j_{ik}:\Omega \to G$ by

3.23 $$f^j_{ik}(s) := \begin{cases} f(r^j_{ik}) & \text{if } s = r^j_{ik} \\ 1_G & \text{elsewhere} , \end{cases}$$

then

3.24 $$(f;\pi) = \prod_{i,j,k} (f^j_{ik};\pi^j_{ik}) ,$$

and these factors $(f^j_{ik};\pi^j_{ik})$ are commutative since they have disjoint cyclic factors if we look at the permutation representation of $G\sim S_n$.

Regarded as elements of subgroups of the form

$$(G_r \times G_{\pi(r)} \times\ldots\times G_{\pi^{k-1}(r)})S_k' \simeq G\sim S_k$$

the n factors $(f_{ik}^j;\pi_{ik}^j)$ are elements of the special form whose centralizers were described in 3.19. Using the same argument as in section 2 to get the centralizer of a general permutation we obtain from 3.19 and 3.24 (Ore [1]):

3.25 <u>If $T(f;\pi) = (a_{ik})$ then the centralizer of $(f;\pi)$ in $G\sim S_n$ is a subgroup conjugate to the centralizer</u>

$$\times_{i,k}(C_{G\sim S_k}(f_{ik}^j;\pi_{ik}^j)\sim S_{a_{ik}})$$

<u>of the special element 3.24 of this type (see 3.20-3.24).</u>

It is easy to check, that this agrees with 3.9, and that for $G = \{1\}$ we get 2.32.

<u>Example</u>. To conclude this section, we consider a subgroup of S_6, which is a faithful permutation representation of $S_3\sim S_2$:

$$(\{1,(12),(13),(23),(123),(132)\} \times \{1,(45),(46),(56),(456),(465)\})$$
$$\cdot\{1,(14)(25)(36)\} \leq S_6 .$$

We would like to describe its conjugacy classes to illustrate 3.7 and give a complete system of representatives of these classes. This we shall do according to the ordering

$$C^1 := \{1\},\quad C^2 := \{(12),(13),(23)\},\quad C^3 := \{(123),(132)\}$$

of the conjugacy classes of S_3.

The order of this group is $3!^2 2! = 72$.

(i) The image of the element $(f;\pi) = (f(1),f(2);\pi) := (1,1;1)$ under the permutation representation described in 2.24/2.25 is of course the identity of S_6 and hence it constitutes a conjugacy class by itself.

The cycleproducts belonging to the two 1-cycles of the permutation $\pi = 1 = (1)(2) \in S_2$ with respect to this mapping $f = e$ are

$$(1) \xrightarrow{f} 1 \in S_3 ,$$
$$(2) \xrightarrow{f} 1 \in S_3 .$$

Therefore $(f;\pi) = (e;1) = (1,1;1)$ is of the type

$$\begin{bmatrix} 2 & 0 \\ 0 & 0 \\ 0 & 0 \end{bmatrix} .$$

(ii) The cycleproducts belonging to the factors of the permutation π of $(f;\pi) := ((12),1;1)$ are

$$(1) \xrightarrow{f} (12)$$
$$(2) \xrightarrow{f} 1 .$$

Hence this element is of type

$$\begin{bmatrix} 1 & 0 \\ 1 & 0 \\ 0 & 0 \end{bmatrix} .$$

The image of this element under the permutation representation is

$$f(1)\sigma f(2)\sigma^{-1} 1^* = (12)\cdot 1 \cdot 1 = (12) \in S_6 .$$

Thus $(12) \in S_6$ is a representing element for this conjugacy class of $S_3 \wr S_2$.

Because of 3.9 the order of this class is

$$72/(3!/1)^1(3!/3)^1 = 6.$$

(iii)((123),1;1) is of type $\begin{bmatrix}1 & 0\\ 0 & 0\\ 1 & 0\end{bmatrix}$. A representative element of this conjugacy class is $(123) \in S_6$ as can be seen analogously to (ii). The order of this class is 4.

(iv) ((12),(12);1) is of type $\begin{bmatrix}0 & 0\\ 2 & 0\\ 0 & 0\end{bmatrix}$. A permutation representing this conjugacy class is (12)(45), the class is of order 9.

(v) $T((123),(123);1) = \begin{bmatrix}0 & 0\\ 0 & 0\\ 2 & 0\end{bmatrix}$. This class is of order 4 and represented by (123)(456).

(vi) $T(1,1;(12)) = \begin{bmatrix}0 & 1\\ 0 & 0\\ 0 & 0\end{bmatrix}$. (1,1;(12)) is mapped onto (14)(25)(36), hence this permutation is a representative element of this conjugacy class of order 6.

(vii) $T((12),(123);1) = \begin{bmatrix}0 & 0\\ 1 & 0\\ 1 & 0\end{bmatrix}$. The image of this class of $S_3 \wr S_2$ contains (12)(456), and its order is 12.

(viii)The cycleproduct belonging to the permutation of ((123), (123);(12)) is

$$(12) \overset{f}{\rightarrow} f(1)f(2) = (132) ,$$

hence

$$T((123),(123);(12)) = \begin{bmatrix}0 & 0\\ 0 & 0\\ 0 & 1\end{bmatrix} .$$

The image under the permutation representation is

$$f(1)\sigma f(2)\sigma^{-1}(12)^* = (123)(456)(14)(25)(36) = (153426) .$$

The order of this class is 12.

(ix) $T((12),(123);(12)) = \begin{bmatrix} 0 & 0 \\ 0 & 1 \\ 0 & 0 \end{bmatrix}$. The image is (15)62634), the order of the class is 18.

These calculations will be of use later on for the evaluation of the character table of $S_3 \sim S_2$. It will be useful for these calculation to observe, that the groups $S_m \sim S_n$ possess two normal divisors of index 2: $S_m \sim S_n{}^+ := S_m \sim S_n \cap A_{mn}$ and $S_m \sim A_n$. Having noticed this we only need to evaluate a small part of the character table and then use the symmetries arising from this fact.
Thus it is helpful to have the representatives of the conjugacy classes in permutational form as well as in the form $(f;\pi)$ as in our example. We see at once that the conjugacy classes described in (i), (iii), (iv), (v) and (ix) are the classes consisting of even permutations and hence these classes form $S_3 \sim S_2{}^+$. Moreover the classes (i), (ii), (iii), (iv), (v) and (vii) are the classes consisting of elements of the form $(f;1)$ and hence they build up the normal divisor $S_3 \sim A_2 = S_3 \sim \{1\} = S_3^* \simeq S_3 \times S_3$.

We have now finished summarizing the group-theoretical results which we need to describe the representation theory of the symmetric and alternating groups as well as of wreath products.

Chapter II

Representations of wreath products

We would like to describe the representations of wreath products, especially of wreath products $G \sim S_n$ with symmetric groups.
A first section contains a preparatory summary of the ordinary representation theory of the symmetric group. The following section contains the representation theory of wreath products of finite groups over an algebraically closed field.

4. The ordinary irreducible representations of the symmetric group

The groundfield is the field $\mathbb{C}$ of complex numbers. Since $\mathbb{C}$ is algebraically closed and of characteristic zero, the number of irreducible $\mathbb{C}$-representations of S_n is equal to the number of conjugacy classes of S_n.

As we have seen in the first section, there is a 1-1-correspondence between the conjugacy classes of S_n and the partitions

$$\alpha = (\alpha_1,\ldots,\alpha_h),\ \alpha_i \in \mathbb{N},\ \alpha_j \geq \alpha_{j+1}\ (1 \leq j \leq h-1),\ \Sigma\, \alpha_i = n$$

of n.

Hence if we can associate with each of these partitions an irreducible $\mathbb{C}$-representation of S_n so that representations associated with different partitions are inequivalent, we have a complete system of irreducible $\mathbb{C}$-representations of S_n.

How this can be done we shall describe now. The procedure using a consideration of primitive idempotents of the group algebra is well known, but a very useful hint to clarify this process by using Meckey's intertwining number theorem we owe to A. J. Coleman (Coleman [1], cf. also Bayar [1], Burrow [1],[2], Gündüzalp [1], Malzan [1], Munkholm [1], and the hint following 2.29 in Robinson [5]).

Together with a partition α we consider the Young-diagram

$$[\alpha]: \quad \begin{array}{ll} \cdot\ \cdot\ \cdots\cdots\cdots\ \cdot & \alpha_1 \text{ nodes} \\ \cdot\ \cdot\ \cdots\ \cdot & \alpha_2 \text{ nodes} \\ \cdots\cdots\cdots & \cdots \\ \cdot\ \cdots\ \cdot & \alpha_h \text{ nodes} \end{array}$$

(cf. section 1) and its <u>first Young-tableau</u>

4.1 $$T_1^\alpha := \begin{array}{l} 1 \quad 2 \ \cdots\cdots\cdots\cdots \ \alpha_1 \\ \alpha_1+1 \quad \alpha_1+2 \ \ldots \ \alpha_1+\alpha_2 \\ \cdots\cdots\cdots\cdots\cdots \\ \cdots\cdots n \end{array}$$

It follows from the results of section 1, that the groups H_1^α and V_1^α of the horizontal and vertical permutations of T_1^α are Young subgroups with the property

4.2 $$H_1^\alpha \cap V_1^\alpha = \{1\} .$$

Let us denote by IH_1^α the identity representation of H_1^α and by AV_1^α the alternating representation of V_1^α, defined by

$$IH_1^\alpha(\pi) := (1) \ , \ \forall \ \pi \in H_1^\alpha \ ,$$

4.3 $$AV_1^\alpha(\pi) := \begin{cases} (1) \ , \ \forall \ \pi \in V_1^{\alpha+} := V_1^\alpha \cap A_n \\ (-1), \ \forall \ \pi \in V_1^\alpha \backslash V_1^{\alpha+} \ . \end{cases}$$

If $IH_1^\alpha \uparrow S_n$ and $AV_1^\alpha \uparrow S_n$ are the representations of S_n induced by IH_1^α and AV_1^α, the following is the crucial theorem:

<u>4.4</u> <u>$IH_1^\alpha \uparrow S_n$ and $AV_1^\alpha \uparrow S_n$ have exactly one irreducible constituent in common, in each case with multiplicity 1.</u>

Proof: The assertion is fulfilled if and only if the inner product $(IH_1^\alpha \uparrow S_n, AV_1^\alpha \uparrow S_n)$ of the corresponding characters $\chi^{IH_1^\alpha \uparrow S_n}$ and $\chi^{AV_1^\alpha \uparrow S_n}$ satisfies

4.5 $$(IH_1^\alpha \uparrow S_n, AV_1^\alpha \uparrow S_n) = 1$$

(see Curtis/Reiner [1], Ex. 31.1). For the inner product is defined by

$$(IH_1^\alpha \uparrow S_n, AV_1^\alpha \uparrow S_n) := \frac{1}{n!} \sum_\pi \chi^{IH_1^\alpha \uparrow S_n}(\pi) \chi^{AV_1^\alpha \uparrow S_n}(\pi^{-1})$$

and hence 4.5 implies the assertion because of the orthogonality relations of the irreducible characters.

To prove 4.5 we notice, that this inner product is the intertwining number of the two induced representations. Hence using Mackey's intertwining number theorem (Curtis/Reiner [1],(44.5)) we obtain

$$(IH_1^\alpha \uparrow S_n, AV_1^\alpha \uparrow S_n) = i(IH_1^\alpha \uparrow S_n, AV_1^\alpha \uparrow S_n)$$

$$= \sum_{H_1^\alpha \pi V_1^\alpha} i(IH_1^\alpha \downarrow H_1^\alpha \cap \pi V_1^\alpha \pi^{-1}, (AV_1^\alpha)^\pi \downarrow H_1^\alpha \cap \pi V_1^\alpha \pi^{-1}),$$

if the sum is taken over a complete system of pairwise different double cosets $H_1^\alpha \pi V_1^\alpha$ of H_1^α and V_1^α in S_n, if "$\downarrow$" means restriction and if we denote by $(AV_1^\alpha)^\pi$ the representation (of $\pi V_1^\alpha \pi^{-1}$) conjugate to AV_1^α and defined by

$$(AV_1^\alpha)^\pi(\pi\pi'\pi^{-1}) := AV_1^\alpha(\pi'), \ \forall\ \pi' \in V_1^\alpha .$$

Now the intersection $H_1^\alpha \cap \pi V_1^\alpha \pi^{-1}$ is obviously a direct product of symmetric subgroups, say

$$H_1^\alpha \cap \pi V_1^\alpha \pi^{-1} = \underset{i,j}{\times} S_{z_{ij}}$$

(cf. 1.31). And therefore the restriction

$$IH_1^\alpha \downarrow H_1^\alpha \cap \pi V_1^\alpha \pi^{-1} = I(H_1^\alpha \cap \pi V_1^\alpha \pi^{-1})$$

is equal to the restriction

$$(AV_1^\alpha)^\pi \downarrow H_1^\alpha \cap \pi V_1^\alpha \pi^{-1} = A(H_1^\alpha \cap \pi V_1^\alpha \pi^{-1})$$

if and only if $z_{ij} \leq 1$, $\forall$ i,j, i.e. if and only if

$$H_1^\alpha \cap \pi V_1^\alpha \pi^{-1} = \{1\} .$$

Hence

$$i(IH_1^\alpha \downarrow H_1^\alpha \cap \pi V_1^\alpha \pi^{-1}, (AV_1^\alpha)^\pi \downarrow H_1^\alpha \cap \pi V_1^\alpha \pi^{-1}) = \begin{cases} 1, \text{ if } H_1^\alpha \cap \pi V_1^\alpha \pi^{-1} = \{1\} \\ 0 \quad \text{otherwise} \end{cases} .$$

From 1.35 we know, that there is exactly one double coset $H_1^\alpha \pi V_1^\alpha$ with the property $H_1^\alpha \cap \pi V_1^\alpha \pi^{-1} = \{1\}$ (namely the double coset $H_1^\alpha V_1^\alpha$, see 4.2). This implies 4.5 and the theorem is proved.

q.e.d.

Let us denote by $[\alpha]$ the equivalence class of this uniquely determined common constituent:

4.6 $$[\alpha] := IH_1^\alpha \uparrow S_n \cap AV_1^\alpha \uparrow S_n .$$

Since with the notation $(1^n) := (1,\dots,1)$ we have

$$IH_1^{(1^n)} = I\{1\} = AV_1^{(n)} ,$$

the induced representation $IH_1^{(1^n)} \uparrow S_n$ as well as $AV_1^{(n)} \uparrow S_n$ is the regular representation RS_n of S_n. Thus

$$IH_1^{(1^n)} \uparrow S_n \cap (AV_1^{(1^n)} \uparrow S_n = AS_n) = RS_n \cap AS_n = AS_n,$$

$$(IH_1^{(n)} \uparrow S_n = IS_n) \cap AV_1^{(n)} \uparrow S_n = IS_n \cap RS_n = IS_n .$$

Using notation 4.6 we get therefrom:

4.7 $$[n] = IS_n\ ,\quad [1^n] = AS_n\ .$$

Let us denote by "#" the outer tensor product multiplication (following the notation of Curtis/Reiner [1], cf. §43) and substitute S_α and $S_{\alpha'}$ for the isomorphic subgroups H_1^α and V_1^α. We obtain from 4.7:

4.8 $$IS_\alpha = [\alpha_1] \# \ldots \# [\alpha_h] = \#_i\ [\alpha_i]\ ,$$
$$AS_{\alpha'} = [1^{\alpha_1'}] \# \ldots \# [1^{\alpha_{h'}'}] = \#_i\ [1^{\alpha_i'}]\ .$$

The induced representations will be denoted as follows:

4.9 $$\prod_i [\alpha_i] = [\alpha_1]\ldots[\alpha_h] := (\#_i[\alpha_i]) \uparrow S_n\ ,$$
$$\prod_i [1^{\alpha_i'}] = [1^{\alpha_1'}]\ldots[1^{\alpha_{h'}'}] := (\#_i[1^{\alpha_i'}]) \uparrow S_n\ .$$

Therefore a second formulation of 4.6 is (cf. Robinson [5],2.29):

4.10 $$[\alpha] = [\alpha_1]\ldots[\alpha_h] \cap [1^{\alpha_1'}]\ldots[1^{\alpha_{h'}'}]\ .$$

Using again the notation of Curtis/Reiner [1] we denote by "⊗" the inner tensor product multiplication. By definition of the alternating representation we have

$$[1^n] \otimes [1^n] = [n]\ .$$

Hence

$$[\alpha'] = IS_{\alpha'} \uparrow S_n \cap AS_\alpha \uparrow S_n = (AS_{\alpha'}\uparrow S_n \cap IS_\alpha\uparrow S_n) \otimes [1^n],$$

thus

<u>4.11</u> $$[\alpha'] = [\alpha] \otimes [1^n] .$$

This means, that the representations $[\alpha]$ and $[\alpha']$ differ only on the odd permutations and there only in the sign.

4.6 characterizes certain irreducible $\mathbb{C}$-representations of S_n, and we would like to show that the system of these representations $[\alpha]$ is a complete system of pairwise inequivalent irreducible $\mathbb{C}$-representations (or ordinary representations as they are also called) of S_n.

To prove this it suffices - as has been said at the beginning of this section - to show, that for $\alpha \neq \beta$ $[\alpha]$ and $[\beta]$ are inequivalent. To prove this we consider minimal left ideals of the group algebra $\mathbb{C}S_n$ of S_n over $\mathbb{C}$, which afford the representations $[\alpha]$.

We know, that the simple two-sided ideal of the group algebra $\mathbb{C}G$ of a finite group G consisting of minimal left ideals affording the irreducible $\mathbb{C}$-representation of G with character ζ is generated by the central and up to a numerical factor idempotent element

4.12 $$\sum_{g \in G} \zeta(g^{-1})g .$$

If this irreducible representation with character ζ is onedimensional, then the simple two-sided ideal is a minimal left ideal itself, since the number of the minimal summands is equal to the dimension of the afforded irreducible representation as is well known. In this case, the element 4.12 is even a primitive idempotent (up to a numerical factor). Applying this to IH_1^α and AV_1^α and setting

4.13 $$\varkappa_1^\alpha := \sum_{\pi\in H_1^\alpha} \pi \,, \quad \gamma_1^\alpha := \sum_{\rho\in V_1^\alpha} \varepsilon_\rho \rho \,,$$

$e_\rho := \pm 1$ if ρ is an even/odd permutation, we obtain:

4.14 IH_1^α is afforded by $CH_1^\alpha\varkappa_1^\alpha$, AV_1^α is afforded by $CV_1^\alpha\gamma_1^\alpha$.

These elements $\varkappa_1^\alpha$ and γ_1^α are up to a numerical factor primitive idempotents of the subalgebras CH_1^α and CV_1^α of CS_n. Regarded as elements of CS_n they are still idempotent up to a numerical factor but of course in general no longer central elements or primitive. The left ideals generated by afford the induced representations:

4.15 $IH_1^\alpha \uparrow S_n$ is afforded by $CS_n\varkappa_1^\alpha \simeq CS_n \otimes_{CH_1^\alpha} CH_1^\alpha\varkappa_1^\alpha$,

$AV_1^\alpha \uparrow S_n$ is afforded by $CS_n\gamma_1^\alpha \simeq CS_n \otimes_{CV_1^\alpha} CV_1^\alpha\gamma_1^\alpha$.

And we would like to show, that the product

4.16 $$e_1^\alpha := \varkappa_1^\alpha\gamma_1^\alpha$$

generates a minimal left ideal of CS_n affording $[\alpha]$.

At first we notice, that because of $H_1^\alpha \cap V_1^\alpha = \{1\}$ the coefficient of 1_{S_n} in e_1^α is 1 and hence

4.17 $$e_1^\alpha \neq 0 \,.$$

4.5 implies, that the C-dimension of $\varkappa_1^\alpha CS_n \gamma_1^\alpha$ is 1:

4.18 $$(\varkappa_1^\alpha CS_n \gamma_1^\alpha : C) = 1.$$

Therefore we obtain with 4.17 <u>John von Neumann's lemma</u> (see Boerner [2], IV, theorem 2.9):

<u>4.19</u> $$\varkappa_1^\alpha CS_n \gamma_1^\alpha = Ce_1^\alpha .$$

Applying this to the special element $\gamma_1^\alpha \varkappa_1^\alpha \in CS_n$ we get

4.20 $$(e_1^\alpha)^2 = \varkappa_1^\alpha \gamma_1^\alpha \varkappa_1^\alpha \gamma_1^\alpha = \varkappa e_1^\alpha,\ \varkappa \in C.$$

It remains to prove, that this complex number $\varkappa$ is unequal to zero and that e_1^α is primitive.

$\varkappa$ can be evaluated by calculating in two ways the trace of the linear transformation of CS_n afforded by the multiplication of CS_n with e_1^α from the right hand side.

At first we assume the basis of CS_n to be adapted to the submodule $CS_n e_1^\alpha$, i.e. that the first $(CS_n e_1^\alpha : C) =: f^\alpha$ basis vectors span $CS_n e_1^\alpha$. With respect to such a basis the multiplication with e_1^α is described by the matrix

4.21 $$\left[\begin{array}{c|c} \begin{matrix} \varkappa & & 0 \\ & \ddots & \\ 0 & & \varkappa \end{matrix} & * \\ \hline 0 & 0 \end{array}\right]$$

on account of 4.20.

If on the other hand we choose the elements π of S_n to be the basis of CS_n, the trace of the multiplication with e_1^α is obviously n!-times the coefficient of 1_{S_n} in e_1^α, which is 1 as we have mentioned above.

Comparing this with 4.21 we obtain

4.22 $$\varkappa = \frac{n!}{(CS_n e_1^\alpha : C)} = \frac{n!}{f^\alpha} .$$

Hence e_1^α is up to the numerical factor $\varkappa^{-1} = f^\alpha/n!$ an idempotent element of CS_n, e_1^α is essentially idempotent.

To show the primitivity of e_1^α we notice, that

4.23 $$CS_n e_1^\alpha = CS_n \varkappa_1^\alpha \gamma_1^\alpha \subseteq (CS_n \varkappa_1^\alpha)(CS_n \gamma_1^\alpha) .$$

The two factors of the right hand side of 4.23 are direct sums of minimal left ideals. Since nonisomorphic minimal left ideals annihilate each other, we obtain from 4.4, that this right hand side of 4.23 is either the ideal $\{0\}$ or a minimal left ideal which affords $[\alpha]$.

Because of $e_1^\alpha \neq 0$ the left ideal $CS_n e_1^\alpha$ is not the zero ideal and hence 4.10 implies:

4.24 <u>$(f^\alpha/n!)e_1^\alpha$ is a primitive idempotent of CS_n and the minimal left ideal $CS_n e_1^\alpha$ generated by e_1^α affords the irreducible C-representation $[\alpha]$ of S_n.</u>

To show the completeness of this system of irreducible ordinary representations $[\alpha]$ of S_n it therefore suffices to prove the following:

4.25 $$\alpha \neq \beta \Rightarrow CS_n e_1^\alpha \not\cong CS_n e_1^\beta .$$

We shall prove this with the aid of a lemma. Before stating this lemma we order the partitions, diagrams and representations according to the row lengths α_i. We say that α respectively $[\alpha]$ <u>precedes</u> β respectively $[\beta]$ (α and β partitions of n), for short: $[\alpha] > [\beta]$, if the first nonvanishing difference $\alpha_i - \beta_i$ is positive. Now the lemma reads as follows:

<u>4.26</u> <u>If $[\alpha] > [\beta]$ and T^α, T^β are Young-tableaux with diagrams $[\alpha]$ and $[\beta]$, then there are at least two symbols which appear in T^α in the same row and in T^β in the same column.</u>

Proof: If this were not the case, the α_1 symbols of the first row of T^α would appear in different columns of T^β so that $\beta_1 \geq \alpha_1$. Since $\alpha > \beta$ this implies $\alpha_1 = \beta_1$.
A vertical permutation - which doesn't disturb the distribution of the symbols in the columns of T^β - transfers these α_1 symbols to the places of the first row of T^β. Disregarding this new first row and using the same argument as above we get $\alpha_2 = \beta_2$ and so on, arriving finally at $\alpha = \beta$ which is a contradiction to the assumption $\alpha > \beta$.

q.e.d.

<u>Proof of 4.25</u>: If $\alpha \neq \beta$ we can assume without restriction that $\alpha > \beta$.
If $\pi \in S_n$, 4.26 implies that in

$$\pi T_1^\alpha := \begin{matrix} \pi(1)\ \pi(2)\ldots\pi(\alpha_1) \\ \ldots\ldots\ldots\ldots\ldots \\ \ldots\pi(n) \end{matrix}$$

there are two symbols, say s and t, appearing in the same row of πT_1^α and in the same column of T_1^β. Their transposition (st) belongs to $\pi H_1^\alpha \pi^{-1}$, the group of the horizontal permutations of πT_1^α as well as to V_1^β. Thus

$$(st)\pi\mathscr{H}_1^\alpha\pi^{-1} = \pi\mathscr{H}_1^\alpha\pi^{-1},\ \mathscr{V}_1^\beta(st) = -\mathscr{V}_1^\beta,$$

from what follows, that

$$e_1^\beta \pi e_1^\alpha \pi^{-1} = -e_1^\beta(st)\pi e_1^\alpha \pi^{-1} = -e_1^\beta \pi e_1^\alpha \pi^{-1}.$$

$$\Rightarrow e_1^\beta \pi e_1^\alpha \pi^{-1} = 0,\ \forall\ \pi\in S_n,\ \Rightarrow e_1^\beta \pi e_1^\alpha = 0,\ \forall\ \pi\in S_n,\ \Rightarrow e_1^\beta x e_1^\alpha = 0, \forall\ x\in \mathbb{C}S_n.$$

And it is well known, that this implies the statement (cf. Boerner [2], III, theorem 3.8).

q.e.d.

We summarize, what we have proved:

4.27 <u>The representations [α] defined by 4.6 form a complete system</u>

$$\{[\alpha] := IS_\alpha\!\uparrow\! S_n \cap AS_{\alpha'}\!\uparrow\! S_n \mid \alpha \text{ partition of } n\}$$

<u>of pairwise inequivalent and irreducible $\mathbb{C}$-representations of S_n.</u>

Thus we have completed the first step towards an explicit description of the ordinary irreducible representations of S_n. The representing matrices themselves were given by A. Young (Young [2]), we shall describe the derivation in detail in a following part. Here we shall only sketch the next steps briefly.

We return to the definition 4.16 of the generating primitive idempotents. To define e_1^α we have used only the first Young-tableau T_1^α with Young-diagram $[\alpha]$. The element

$$e_1^\alpha = \varkappa_1^\alpha \gamma_1^\alpha$$

constructed with the aid of the groups H_1^α and V_1^α of the horizontal and vertical permutations of T_1^α generates a minimal left ideal out of the simple two-sided ideal in CS_n to which the irreducible ordinary representation $[\alpha]$ corresponds.

This simple two-sided ideal is a direct sum of $f^\alpha = (CS_n e_1^\alpha : C)$ minimal left ideals which are isomorphic to $CS_n e_1^\alpha$. Thus we ask for elements generating the remaining minimal left ideals out of this simple two-sided ideal.

Presumably some of the elements

$$e_i^\alpha := \varkappa_i^\alpha \gamma_i^\alpha$$

constructed analogously to e_1^α but with the aid of other tableaux T_i^α generate these left ideals.

This is actually the case (that $CS_n e_i^\alpha \simeq CS_n e_1^\alpha$ is trivial by definition of $[\alpha]$). Obviously there are $n!$ different tableaux with Young-diagram $[\alpha]$, namely the tableaux πT_1^α, $\pi \in S_n$. We pick out some of them, the so-called <u>standard-tableaux</u>, characterized by the property, that in such a tableau the symbols in each row and in each column are in increasing order, i.e. that the symbol in the position (i,j) (i-th row, j-th column) precedes that in the (k,l)-position if $i \leq k$ and $j \leq l$. E.g.

$$\begin{matrix} 1\,2\,3 & 1\,2\,4 & 1\,2\,5 & 1\,3\,4 & 1\,3\,5 \\ 4\,5 & 3\,5 & 3\,4 & 2\,5 & 2\,4 \end{matrix}$$

are all the standard-tableaux with Young-diagram [3,2].

Let us denote by f_α the number of standard-tableaux with Young-diagram $[\alpha]$. It will appear, that $f_\alpha = f^\alpha = (\mathbb{C}S_n e_1^\alpha : \mathbb{C})$.

We arrange the standard-tableaux T_i^α in <u>dictionary order</u>, i.e. $i<j$ if the first nonvanishing difference of symbols located at the same place in T_i^α and T_j^α (comparing the symbols in a row from the left to the right and the rows downwards) is negative. The above arrangement of the standard tableaux with diagram [3,2] provides an example of this ordering.

Let H_i^α and V_i^α be the groups of the horizontal permutations and of the vertical permutations of T_i^α, let

4.28 $$\varkappa_i^\alpha := \sum_{\pi\in H_i^\alpha} \pi \quad , \quad \mathscr{V}_i^\alpha := \sum_{\rho\in V_i^\alpha} \varepsilon_\rho \rho \ ,$$

and

4.29 $$e_i^\alpha := \varkappa_i^\alpha \mathscr{V}_i^\alpha = \sum_{\pi\in H_i^\alpha, \rho\in \mathscr{V}_i^\alpha} \varepsilon_\rho \pi\rho, \ 1 \leq i \leq f_\alpha .$$

We would like to show, that

4.30 $$\sum_{\alpha, i} \mathbb{C}S_n e_i^\alpha$$

is a direct sum and equal to $\mathbb{C}S_n$. This would imply $f^\alpha = f_\alpha$ and that $\bigoplus_i \mathbb{C}S_n e_i^\alpha$ is the simple two-sided ideal of $\mathbb{C}S_n$, whose summands

$CS_n e_i^\alpha$ afford $[\alpha]$.

The first step towards a proof of 4.30 is the proof of

<u>4.31</u> $$i < j \Rightarrow e_j^\alpha e_i^\alpha = 0.$$

And this will be shown with the aid of the following lemma:

<u>4.32</u> <u>If $i<j$, then there are at least two symbols appearing in the same row of T_i^α and in the same column of T_j^α.</u>

Proof: Let (k,l) be the place of the first node of $[\alpha]$ replaced by different symbols, say by s resp. t in T_i^α resp. T_j^α. $i<j$ implies $s<t$. We ask for the place (m,o), where s appears in T_j^α.

On account of the standardness of T_j^α, t precedes all the symbols located at places (p,q) with $p \geq k$ and $q \geq l$. Since (k,l) is the first place occupied by different symbols, this implies $m>k$, $o<l$. Thus the place (k,o) is occupied by the same symbol, say by r, in T_i^α as well as in T_j^α. Hence r and s fulfil the statement.

q.e.d.

Denoting by r and s two symbols appearing in the same row of T_i^α and in the same column of T_j^α, we have for their transposition (rs):

$$(rs) \in H_i^\alpha \cap V_j^\alpha \Rightarrow e_j^\alpha e_i^\alpha = -e_j^\alpha (rs) e_i^\alpha = -e_j^\alpha e_i^\alpha \Rightarrow e_j^\alpha e_i^\alpha = 0 .$$

This proves 4.31.

To proceed with the proof of that 4.30 is a direct sum and equal to CS_n, we consider an equation

$$x_1 e_1^\alpha + \ldots + x_{f_\alpha} e_{f_\alpha}^\alpha = 0$$

($x_i \in CS_n$). If we multiply by e_1^α from the right hand side, since 4.31 is valid, we obtain $\kappa x_1 e_1^\alpha = 0$, hence $x_1 e_1^\alpha = 0$. Then we multiply by e_2^α and get $x_2 e_2^\alpha = 0$ and so on. This proves, that the sum 4.30 is a direct sum.

A combinatorial consideration (see Boerner [2], IV, § 7) shows that

4.33 $$\sum_\alpha (f_\alpha)^2 = n! \ .$$

Since 4.30 is a direct sum we have $f^\alpha \geq f_\alpha$, together with 4.33 we conclude, that $f^\alpha = f_\alpha$ and this completes the proof, so that the following is valid:

<u>4.34</u> $$CS_n = \bigoplus_\alpha \bigoplus_{i=1}^{f^\alpha} CS_n e_i^\alpha \ .$$

Hence

4.35 $$\bigoplus_{i=1}^{f^\alpha} CS_n e_i^\alpha$$

is the two-sided ideal to which [α] belongs.

The theorem of Wedderburn says: $\bigoplus_i CS_n e_i^\alpha$ is isomorphic to the ring of ($f^\alpha \times f^\alpha$)-matrices over C. Hence all what remains to construct representing matrices themselves is to find a basis of elements e_{ik}^α of $\bigoplus_i CS_n e_i^\alpha$ satisfying

4.36 $$e_{ij}^\alpha e_{kl}^\alpha = \delta_{jk} e_{il}^\alpha$$

(δ_{jk} the Kronecker-symbol: $\delta_{jk}=0$ if $j \neq k$, $=1$, if $j=k$).

Thus the matrices $(d_{ik}^\alpha(\pi))$ built from the coefficients of

4.37 $$\pi = \sum_{i,k=1}^{f^\alpha} d_{ik}^\alpha(\pi) e_{ik}^\alpha$$

are the elements of $[\alpha]$. For this the reader is referred to Boerner [1], [2]. Corresponding to three ways of choosing the basis elements e^{α}_{ik} there are three forms of the representing matrices: Young's natural form ($d^{\alpha}_{ik} \in Z$), Young's seminormal form ($d^{\alpha}_{ik} \in Q$) and Young's orthogonal form ($d^{\alpha}_{ik} \in R$, the matrices are orthogonal). The natural form is derived in Boerner [1], the orthogonal and seminormal form in Boerner [2]. We shall describe the seminormal form now.

To get the matrices representing elements of the subgroup $S_{n-1} \leq S_n$ (consisting of the permutations fixing the symbol $n \in \Omega$) in reduced form, i.e. with the matrices of the irreducible constituents along the main diagonal and zeros elsewhere, we choose an ordering of the standard-tableaux with respect to this symbol n: to obtain the last letter sequence $T^1_{\alpha}, \ldots, T^{f^{\alpha}}_{\alpha}$ of the standard-tableaux with Young-diagram $[\alpha]$, we take at first the standard-tableaux containing n in the last row, then the standard-tableaux containing n in the last but one row and so on. Then we order the standard-tableaux containing n in the same row with respect to n-1 and so on. E.g.

4.38

```
1 2 3    1 2 4    133 4    1 2 5    1 3 5
4 5      3 5      2 5      3 4      2 4
```

is the last letter sequence of the standard-tableaux with Young-diagram [3,2]. To distinguish between the last letter sequence and the dictionary ordering we have exchanged the indices: T^i_{α} may be different from T_i^α.

Before we can formulate the theorem we have to introduce the axial distance $d_\alpha^i(r,s)$ of two symbols r and s in T_α^i: if r is located at (i_r,j_r) and s at (i_s,j_s), we define:

4.39 $$d_\alpha^i(r,s) := (j_r-j_s) + (i_s-i_r) \; .$$

Thus $d_\alpha^i(r,s)$ is the number of steps we need to come from r to s, if steps to the left and downwards are counted positively and steps to the right and upwards are counted negatively.

Now the theorem describing Young's seminormal form of $[\alpha]$ reads as follows:

4.40 If $T_\alpha^1, \ldots, T_\alpha^{f^\alpha}$ is the last letter sequence of the standard-tableaux with Young-diagram $[\alpha]$, then from the matrices $(d_{ik}^\alpha(t,t+1))$ representing the transpositions $(t,t+1)$ $(1 \leq t \leq n-1)$ can be built a representation equivalent to $[\alpha]$ if we set

(i) $d_{ii}^\alpha(t,t+1) := \pm 1$, if T_α^i contains t and t+1 in the same row/column,

(ii) and for the submatrix

$$\begin{bmatrix} d_{ii}^\alpha(t,t+1) & d_{ij}^\alpha(t,t+1) \\ d_{ji}^\alpha(t,t+1) & d_{jj}^\alpha(t,t+1) \end{bmatrix} := \begin{bmatrix} -d_\alpha^i(t,t+1)^{-1} & 1-d_\alpha^i(t,t+1)^{-2} \\ 1 & d_\alpha^i(t,t+1)^{-1} \end{bmatrix}$$

if $T_\alpha^i = (t,t+1)T_\alpha^j$ and $i<j$,

(iii) $d_{ij}^\alpha(t,t+1) := 0$ for all the other entries.

Later on we shall use this theorem for the evaluation of examples of representations of wreath products. For this we shall also use some results about the ordinary irreducible characters of the symmetric group which we summarize now.

A tool of great utility is the theorem, that the character ζ^α of $[\alpha]$ can be written in the <u>determinantal form</u>

<u>4.41</u> $$[\alpha] = |[\alpha_i+j-i]|$$

as a linear combination (with rational integral coefficients) of characters induced by identity representations of Young subgroups. 4.41 has to be understood as follows:

4.42 $$\zeta^\alpha = \sum_{\pi\in S_h} \varepsilon_\pi \chi^{\Pi[\alpha_i+\pi(i)-i]},$$

if $\chi^{\Pi[\beta_i]}$ denotes the character of $^-$ $[\beta_i]$ (see 4.9), if $\beta_i \geq 0$, $\forall$ i, and if we set $\chi^{\Pi[\beta_i]} = 0$, if one $\beta_i < 0$, and if we set in the determinantal expression 4.41:

$$[0] := 1,\ [m] := 0, \text{ if } m<0 .$$

For example

$$[3,1^2] = \begin{vmatrix} [3] & [4] & [5] \\ 1 & [1] & [2] \\ 0 & 1 & [1] \end{vmatrix} = [3][1][1] - [3][2] - [4][1] + [5] ,$$

what means, that

$$\zeta^{(3,1^2)} = \chi^{[3][1][1]} - \chi^{[3][2]} - \chi^{[4][1]} + \chi^{[5]} .$$

Above all it is remarkable, that 4.41 like 4.4 can be proved using the results about double cosets mentioned in section 1 and Mackey's intertwining number theorem (see Coleman [1], cf. also Gündüzalp [1]) so that we can get a homogeneous and lucid derivation of these fundamental and classical results following Coleman's hint to use Mackey's theorem.

A corollary of 4.41 is

4.42 $$P\pi = (n) \Rightarrow \zeta^{\alpha}(\pi) = \begin{cases} (-1)^r, & \text{if } [\alpha]=[n-r,1^r], 0\leq r\leq n-1 \\ 0 & \text{elsewhere} \end{cases} .$$

For the only Young subgroup containing n-cycles is S_n itself. Hence an n-cycle π has a nonvanishing character value $\zeta^{\alpha}(\pi)$ at most when $[n]$ occurs in the determinantal expression 4.41, and this is obviously the case only if $[\alpha]$ is of the form $[n-r,1^r]$. In this case we have

$$[n-r,1^r] = \begin{vmatrix} [n-r] & [n-r+1] & \dots & [n-1] & [n] \\ 1 & [1] & \dots & [r-1] & [r] \\ \dots & \dots & \dots & \dots & \dots \\ 0 & 0 & \dots & 1 & [1] \end{vmatrix}$$

$$= (-1)^r[n] + \underbrace{\dots\dots\dots\dots}_{\text{each summand} \neq [n]} ,$$

hence 4.42 is valid.

To prove 4.41 we introduce the concept of a hook, whose importance was first noted by T. Nakayama (Nakayama [1]).

With the aid of this concept a very simple recursion formula for ζ^{α} can be formulated and a very simple equation for f^{α} can be given.

A <u>hook</u> we call each Γ-shaped arrangement H^{α}_{ij} of nodes out of a Young-diagram $[\alpha]$ which consists of the (i,j)-node, the <u>corner</u> of the hook, as well as of the (i,k)-nodes, $k>j$, which form the <u>arm</u> of the hook, and the (l,j)-nodes, $l>i$, which form the <u>leg</u> of the hook H^{α}_{ij}. The (i,α_i)-node is called the <u>hand</u> of the hook, the (α'_j,j)-node is called the <u>foot</u> of H^{α}_{ij}:

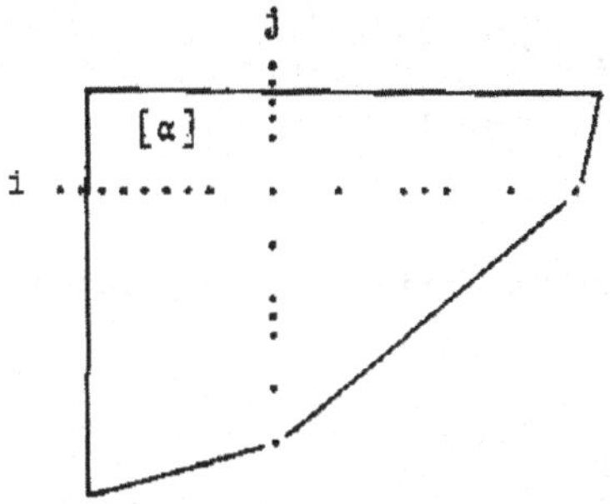

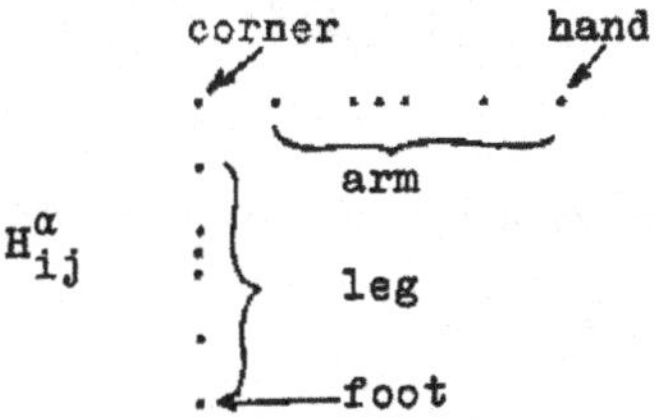

The number

4.43 $$h^{\alpha}_{ij} := \alpha_i - j + \alpha'_j - i + 1$$

of nodes the hook H^{α}_{ij} consists of is called the <u>length</u> of the hook. The number

4.44 $$l^{\alpha}_{ij} := \alpha'_j - i$$

is called the <u>leg-length</u> of H^{α}_{ij}.

To H^{α}_{ij} corresponds the part of length h^{α}_{ij} of the <u>rim</u> of $[\alpha]$ consisting of the nodes between the hand and the arm of H^{α}_{ij} inclusive. E.g.

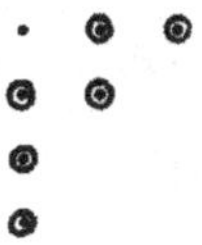

where the encircled nodes indicate the part of the rim which corresponds to the hook $H_{11}^{(3,2,1^2)} \subseteq [3,2,1^2]$.

This <u>associated part of the rim</u> will be denoted by

$$R_{ij}^{\alpha} .$$

And it is important, that the result $[\alpha]\backslash R_{ij}^{\alpha}$ of removing R_{ij}^{α} from $[\alpha]$ is a Young-diagram again or equal to $[0]$ (what is sometimes called the <u>zero-diagram</u>). E.g.

$[3,2,1^2]\backslash R_{11}^{(3,2,1^2)} =$  $= [1]$.

Using this notation we can formulate the following two very important formulae:

<u>4.45</u> (<u>"Murnaghan-Nakayama-formula"</u>)

<u>If $\pi \in S_n$ is of type $T\pi = (a_1,\ldots,a_n)$ so that $a_k \neq 0$, and $\pi^* \in S_{n-k}$ is of type $T\pi^* = (a_1,\ldots,a_{k-1},a_k-1,a_{k+1},\ldots,a_n)$, then</u>

$$\zeta^{\alpha}(\pi) = \sum_{i,j:h_{ij}^{\alpha}=k} (-1)^{l_{ij}^{\alpha}} \zeta^{[\alpha]\backslash R_{ij}^{\alpha}}(\pi^*) ,$$

<u>if we set $\zeta^{[0]} := 1$.</u>

This is the recursion formula for the ordinary irreducible characters of S_n. The formula for the dimension of $[\alpha]$ we owe to Frame, Robinson and Thrall (Frame/Robinson/Thrall [1]), and it is of fascinating simplicity:

<u>4.46</u> <u>The dimension of [α] is the quotient of n! and the product of all the hook-lengths:</u>

$$f^{\alpha} = n!/\prod_{i,j} h^{\alpha}_{ij} .$$

4.45 implies, that $\zeta^{\alpha}(\pi) = 0$ if π contains a k-cycle but [α] contains no k-hook, and it implies 4.42.

It should be observed, that the order of removing two parts of the rim associated with hooks is immaterial, the resulting diagram will be the same in both cases.

Now we have mentioned the most important results about the ordinary irreducible representations. Certain analogues concerning reducible representations can be formulated, which will be of use later on.

We consider the representations

4.47 $$[\alpha][\beta] := [\alpha]\#[\beta] \uparrow S_{m+n} ,$$

[α] an irreducible representation of S_m, [β] an irreducible representation of S_n.

There is an analogue to 4.40 which describes the representing matrices of [α][β].

To formulate this we have to generalize the notation of the Young-diagram. To do this we consider a diagram [α] and a diagram [β] which can be superimposed upon [α], upper left hand corner upon upper left hand corner such that [β] is contained entirely within [α]:

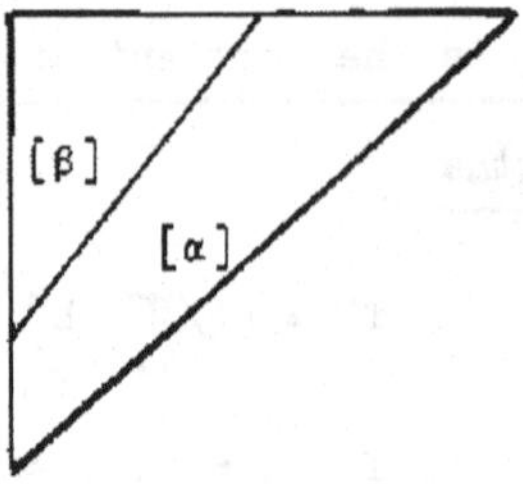

The residuum of [α] not covered by [β] is called a <u>skew-diagram</u> and denoted by

$$[\alpha] - [\beta] .$$

E.g.

[7,6,3,1] - [7,2,1] = 

= [6,3,1] - [2,1]

([α]-[β] is by no means uniquely determined in terms of [α] and [β]).

Special cases of skew diagrams consist of two disjoint diagrams:

[α;β] := 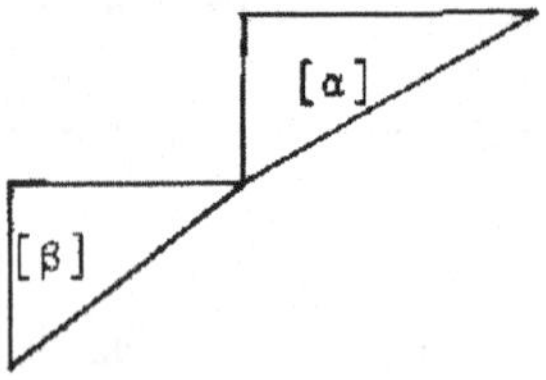

E.g.

$[4,3,2,1] - [2^2] = [2,1;2,1] =$ 

We shall consider only these special cases of skew diagrams.

Tableaux and standard-tableaux $T^i_{\alpha;\beta}$ are defined as before.

The number of standard-tableaux with Young-diagram $[\alpha;\beta]$ is obviously

4.48 $$\binom{m+n}{m} f^{\alpha} f^{\beta} = \frac{(m+n)!}{m!n!} f^{\alpha} f^{\beta} ,$$

i.e. equal to the dimension of $[\alpha][\beta]$. The theorem analogous to 4.40 reads as follows (Robinson [5], 3.1):

4.49 If $T^1_{\alpha;\beta},\ldots,T^{f^{\alpha;\beta}}_{\alpha;\beta}$ ($f^{\alpha;\beta} := (m+n)!f^{\alpha}f^{\beta}/m!n!$) is the last letter sequence of the standard-tableaux with skew diagram $[\alpha;\beta]$, then we can build a representation equivalent to $[\alpha][\beta]$, if we take for the entries of the matrices $(d^{\alpha;\beta}_{ij}(t,t+1))$ representing the transpositions $(t,t+1)$:

(i) $d^{\alpha;\beta}_{ii}(t,t+1) = \pm 1$, if t and $t+1$ occur in the same row/column of $T^i_{\alpha;\beta}$,

(ii) $$\begin{bmatrix} d^{\alpha;\beta}_{ii}(t,t+1) & d^{\alpha;\beta}_{ij}(t,t+1) \\ d^{\alpha;\beta}_{ji}(t,t+1) & d^{\alpha;\beta}_{jj}(t,t+1) \end{bmatrix} := \begin{bmatrix} -d^i_{\alpha;\beta}(t,t+1)^{-1} & 1-d^i_{\alpha;\beta}(t,t+1)^{-2} \\ 1 & d^i_{\alpha;\beta}(t,t+1)^{-1} \end{bmatrix}$$
$$\text{or} \quad := \begin{bmatrix} 0 & 1 \\ 1 & 0 \end{bmatrix}$$
if t and $t+1$ occur in $T^i_{\alpha;\beta}$ in the same diagram constituent or not.

(iii) $d^{\alpha;\beta}_{ij}(t,t+1) = 0$ for all the other entries.

The formula analogous to the Murnaghan-Nakayama-formula reads as follows (Osima [1]):

4.50 If $\pi \in S_{m+n}$ is of type $T\pi = (a_1,\ldots,a_{m+n})$ with $a_k \neq 0$ and if

$\pi^* \in S_{m+n-k}$ is of type $T\pi^*=(a_1,\ldots,a_{k-1},a_k-1,a_{k+1},\ldots,a_{m+n})$, then we have for the value $\chi^{\alpha;\beta}(\pi)$ of the character of $[\alpha][\beta]$ on π:

$$\chi^{\alpha;\beta}(\pi) = \sum_{i,j:h_{ij}^{\alpha;\beta}=k} (-1)^{l_{ij}^{\alpha;\beta}} \chi^{[\alpha;\beta]\backslash R_{ij}^{\alpha;\beta}}(\pi^*),$$

if we set $\chi^{[0]} := 1$.

The following theorem (stated by Littlewood and Richardson and proved first by Robinson) describes, how we can get the irreducible constituents of $[\alpha][\beta]$:

4.51 ("Littlewood-Richardson-rule")

The irreducible constituents of $[\alpha][\beta]$ are exactly the representations $[\gamma]$ of S_{m+n}, whose diagrams arise by adding the nodes of $[\beta]$ to $[\alpha]$ according to the following rules:

(i) To $[\alpha]$ add the nodes of the first row of $[\beta]$. These may be added to one row or divided into subsets preserving their order and added to different rows, the first subset to one row of $[\alpha]$, the second to a subsequent row, the third to a subsequent to this and so on. After the additions the resulting diagram may not contain two added nodes in the same column.

(ii) Next add the second row of $[\beta]$ according to the same rules followed by the remaining rows in succession and such that each node of $[\beta]$ appears in a later row of the compound diagram $[\gamma]$ than that node immediately above it in $[\beta]$.

A special and very important case is the branching theorem for the irreducible ordinary representations of S_n:

4.52
$$[\alpha] \downarrow S_{n-1} = \sum_{i,j:h^{\alpha}_{ij}=1} ([\alpha]\backslash R^{\alpha}_{ij}) .$$

It is easily to prove this using 4.51 and the reciprocity theorem of Frobenius. For 4.51 says, that the constituents of $[\alpha] \uparrow S_{n+1} = [\alpha][1]$ are exactly the representations $[\gamma]$ of S_{n+1}, whose diagrams arise by adding a node to $[\alpha]$. Thus by Frobenius' reciprocity theorem the diagrams of the constituents of $[\alpha] \downarrow S_{n-1}$ arise by subtracting a node in all the possible ways.

Concluding this section we would like to consider the connection between the irreducible $\mathbb{C}$-representations of S_n and A_n.
A first remark follows immediately from 4.11:

4.53
$$[\alpha] \downarrow A_n = [\alpha'] \downarrow A_n .$$

Thus $[\alpha]$ and $[\alpha']$ are associated representations in the sense of Clifford's theory of representations of groups with normal divisors (Clifford [1], cf. also Boerner [2], III, § 13). But from Clifford's theory we get much more:

4.54 (i) $\alpha \neq \alpha' \Rightarrow [\alpha] \downarrow A_n = [\alpha'] \downarrow A_n$ is irreducible.

(ii) $\alpha = \alpha' \Rightarrow ([\alpha]=[\alpha']) \downarrow A_n = [\alpha]^+ + [\alpha]^-$ with two irreducible and conjugate representations $[\alpha]^{\pm}$ of A_n (i.e. $[\alpha]^{+(a)}$ is equivalent to $[\alpha]^-$, $\forall\, a \in S_n$, if $[\alpha]^{+(a)}(a\pi a^{-1}) := [\alpha]^+(\pi)$, $\forall\, \pi \in A_n$).

Hence for the characters we may use the Murnaghan-Nakayama-formula again in the case $\alpha \neq \alpha'$. If $\alpha = \alpha'$ difficulties arise also because of the splitting of the S_n-classes. But we have (see Boerner [2]) a result of Frobenius:

4.55 ("Frobenius' theorem")

If $\alpha = \alpha'$ the class with the partition

$$\beta := (h_{11}^{\alpha}, h_{22}^{\alpha}, \ldots, h_{kk}^{\alpha})$$

is a splitting S_n-class $C^{\beta} = C^{\beta+} \cup C^{\beta-}$. On this class ζ^{α} has the value

$$\zeta_{\beta}^{\alpha} = (-1)^{(n-k)/2},$$

and the values of $\zeta^{\alpha\pm}$ on $C^{\beta\pm}$ are

$$\zeta_{\beta+}^{\alpha\pm} = \tfrac{1}{2}(\zeta_{\beta}^{\alpha} \pm \sqrt{\zeta_{\beta}^{\alpha} \textstyle\prod_i h_{ii}^{\alpha}}\ )$$

$$\zeta_{\beta-}^{\alpha\pm} = \tfrac{1}{2}(\zeta_{\beta}^{\alpha} \mp \sqrt{\zeta_{\beta}^{\alpha} \textstyle\prod_i h_{ii}^{\alpha}}\ ),$$

if we denote the irreducible constituents of $[\alpha] \downarrow A_n$ in this way. On all the other classes with partitions $\gamma \neq \beta$, we have

$$\zeta_{\gamma}^{\alpha\pm} = \zeta_{\gamma}^{\alpha}/2 .$$

We recall the considerations of the first section concerning the ambivalency of alternating groups (cf. 1.25) and take into account, that the ambivalency of a group is equivalent to the reality of its ordinary irreducible characters. Hence the tables of $A_1 = A_2 = \{1\}$, A_5, A_6, A_{10} and A_{14} are the only character tables of alternating groups containing only real entries, while e.g.

A_3	(3,0,0)	$(0,0,1)^+$	$(0,0,1)^-$
[3]	1	1	1
$[2,1]^+$	1	$(-1+i\sqrt{3})/2$	$(-1-i\sqrt{3})/2$
$[2,1]^-$	1	$(-1-i\sqrt{3})/2$	$(-1+i\sqrt{3})/2$

is the character table of A_3. The table of S_n is real for each n. It should be mentioned, that S_n, A_n and the wreath products of the form $C_p \sim S_n$ (p a prime, $n \geq 3$) are Nagao-groups, i.e. they are the only groups with these character tables (see Nagao [1], Oyama [1], Yokonuma [1]).

For the construction of the representing matrices see Puttaswamaiah [1] and Puttaswamaiah/Robinson [1].

These theorems contain the results we need from the ordinary representation theory of S_n and A_n. Below we shall summarize the results of the modular theory. The results from the ordinary theory suffice to give a detailed description of the ordinary representation theory of $G \sim S_n$. But before deriving this theory, we would like to consider the more general case $G \searrow H$, the groundfield will be assumed to be algebraically closed.

5. Representations of wreath products

Let K denote an algebraically closed field, G a finite group and H a permutation group of degree n, i.e. a subgroup of S_n, the symmetric group on $\Omega = \{1,\dots,n\}$.

We would like to derive, how the irreducible K-representations of $G\sim H$ can be constructed.

For K = ℂ, W. Specht has done this in 1933 (Specht [2]) after having treated the special case $G\sim S_n$ (Specht [1]). His results can be generalized to groundfields K of any characteristic, if K is assumed to be algebraically closed. Using the theory of representations of groups with normal divisors given by A.H. Clifford in 1937 (Clifford [1]) the derivation of the desired results can be shortened considerably (Kerber [2],[4]). We describe this now.

We apply Clifford's theory to the normal divisor

$$G^* = G_1 \times \dots \times G_n \trianglelefteq G\sim H ,$$

the basis group of $G\sim H$, which is a direct product of n subgroups G_i isomorphic to G:

$$G \simeq G_i := \{(f;1_H) \mid f:\Omega \to G,\ f(j)=1_G,\ \forall\ j\neq i\} \leq G\sim H$$

(cf. section 2).

Since the groundfield K is assumed to be algebraically closed, the irreducible K-representations of G^* are exactly the outer tensor products

5.1 $$F^* := F_1 \# \ldots \# F_n$$

of irreducible K-representations F_i of G with the representing matrices

5.2 $$F^*(f;1_H) := F_1(f(1)) \times \ldots \times F_n(f(n)) \quad \text{(Kronecker product).}$$

At first we have to derive the <u>inertia group</u> $G \sim H_{F^*}$ of this representation F^*, which is defined by

5.3 $$G \sim H_{F^*} := \{(f;\pi) \mid F^{*(f;\pi)} \sim F^*\}$$

("$\sim$" indicates equivalency, $F^{*(f;\pi)}(f';1_H) := F^*(f;\pi)^{-1}(f';1_H)(f;\pi)$ $= F^*(f^{-1}_{\pi^{-1}} f'_{\pi^{-1}} f_{\pi^{-1}};1_H) = F^*((f^{-1}f'f)_{\pi^{-1}};1_H))$.

Since $G^* \leq G \sim H_{F^*}$ this group is obviously a product

5.4 $$G \sim H_{F^*} = G^* H'_{F^*}$$

of G^* with a subgroup H'_{F^*} of the complement H' of G^*. H'_{F^*} will be called the <u>inertia factor</u> of F^*:

5.5 $$H'_{F^*} = \{(e;\pi) \mid F^{*(e;\pi)} \sim F^*\} .$$

We notice, that

5.6 $$F^{*(e;\pi)}(f;1_H) = F^*(e;\pi)^{-1}(f;1_H)(e;\pi) = F^*(f_{\pi^{-1}};1_H) .$$

To describe the inertia factor explicitly we distinguish the irreducible K-representations 5.1 with respect to their type:

<u>5.7 Def.:</u> Let $F^1,\ldots,F^r$ be a fixed arrangement of the r pairwise inequivalent K-representations of G.

We call $F^* = F_1 \# \ldots \# F_n$ to be of <u>type</u> $(n)=(n_1,\ldots,n_r)$

(with respect to the above arrangement), if n_j is the number of factors F_i of F^* equivalent to F^j.

Let F^* be of type (n) and let S_{n_j} be the subgroup of S_n ($\geq H$) consisting of the elements permuting exactly the n_j indices of the n_j factors F_i of F^* which are equivalent to F^j. We set

5.8 $\quad S'_{(n)} := S'_{n_1} \times \ldots \times S'_{n_r}$ with $S'_{n_j} := \{(e;\pi) \mid \pi \in S_{n_j}\}$.

We would like to prove, what is suggested by 5.6:

5.9
$$H'_{F^*} = H' \cap S'_{(n)} .$$

Proof: 5.2 and 5.6 imply

$$F^{*(e;\pi)}(f;1_H) = F_1(f(\pi(1)))\times\ldots\times F_n(f(\pi(n))) .$$

The question is, for which π this representation is equivalent to F^*. Since F^* as well as $F^{*(e;\pi)}$ are irreducible representations, we can use a character-theoretical argument.

(i) If $\pi \in H \cap S_{(n)}$, then the trace of $F_i(f(\pi(i)))$ is equal to the trace of $F_{\pi(i)}(f(\pi(i)))$. Thus

$$\mathrm{tr}\, F^*(f;1_H) = \mathrm{tr}\, F^{*(e;\pi)}(f;1_H),\ \forall f, \Rightarrow F^* \sim F^{*(e;\pi)}$$

$$\Rightarrow H' \cap S'_{(n)} \leq H'_{F^*} .$$

(ii) If the other way round $(e;\pi)$ is an element of the inertia factor, we have

$$\mathrm{tr}\, F^*(f;1_H) = \mathrm{tr}\, F^{*(e;\pi)}(f;1_H) = \mathrm{tr}\, F^*(f_{\pi^{-1}};1_H),\ \forall f.$$

If we choose $(f;1_H) \in G_i$ such that $f(j) = 1_G$, $\forall\, j \neq i$, then in this special case

$$(\prod_{j\neq i} f^j)\operatorname{tr} F_i(f(i)) = (\prod_{j\neq\pi^{-1}(i)} f^j)\operatorname{tr} F_{\pi^{-1}(i)}(f(i)) \qquad (1)$$

with the dimensions f^j of the factors F_j of F^*. And this is valid for each $f(i) \in G$.

Let us consider the associated Brauer characters (if charK=p) resp. the traces (if charK=0) φ^i. We are allowed to simplify (1) so that we obtain

$$\varphi^i(f(i))f^{\pi^{-1}(i)} = \varphi^{\pi^{-1}(i)}(f(i))f^i, \ \forall\ f(i) \in G.$$

Since the rows of the character tables and the Brauer characters of irreducible modular representations are linearly independent, we obtain

$$\varphi^i = \varphi^{\pi^{-1}(i)} \Rightarrow F_i \sim F_{\pi^{-1}(i)} \Rightarrow \pi \in S_{(n)} \Rightarrow H'_{F^*} \leq H'\cap S'_{(n)}.$$

q.e.d.

From this we obtain for the inertia group:

<u>5.10</u> $$G\sim H_{F^*} = G^*(H\cap S_{(n)})' = G\sim(H\cap S_{(n)}) .$$

Following Clifford's theory we now have to extend F^* to a representation of its inertia group.

It is not possible to extend a linear representation of a normal divisor to a linear representation of the inertia group in general. We can often extend it to a projective representation. But in our case here, where G^*, the normal divisor, is a semidirect factor of the inertia group, we are fortunately able to extend F^* to a linear representation of $G\sim H_{F^*}$. How this can be done, Specht has shown (Specht [1]).

We can assume without restriction, that equivalent factors F_i of F^* are not only equivalent, but even equal:

5.11 $$F_j \sim F_k \Rightarrow F_j(f(i)) = F_k(f(i)), \ \forall f(i) \in G .$$

If now

5.12 $$F^*(f;1_H) = (f^1_{\alpha_1\beta_1}(f(1))\dots f^n_{\alpha_n\beta_n}(f(n)))$$

is the matrix representing $(f;1_H)$, then we set for $\pi \in H\cap S_{(n)}$:

5.13 $$\tilde{F}^*(f;\pi) := (f^1_{\alpha_1\beta_{\pi^{-1}(1)}}(f(1))\dots f^n_{\alpha_n\beta_{\pi^{-1}(n)}}(f(n))) ,$$

and it is easy to verify, that these matrices form a representation $\tilde{F}^*$ of $G\sim H_{F^*}$. Since

5.14 $$\tilde{F}^* \downarrow G^* = F^* ,$$

this representation is irreducible.

Now let F'' be an irreducible K-representation of $H\cap S_{(n)}$ (if $H'\cap S_{(n)}$ is the inertia factor of F^*) and F' according to

5.15 $$F'(f;\pi) := F''(\pi)$$

the corresponding representation of the inertia group. Multiplying these two representations together, the result

5.16 $$\tilde{F}^* \otimes F'$$

with the representing matrices

5.17 $$(\tilde{F}^* \otimes F')(f;\pi) := \tilde{F}^*(f;\pi)\times F'(f;\pi)$$

is an irreducible K-representation of $G\sim H_{F^*}$, as can be seen with the aid of Clifford's theory.

The most important result is, that the representation induced by

5.16 is irreducible:

5.18 $F := (\tilde{F}^* \otimes F') \uparrow G \setminus H$ is irreducible and every irreducible K-representation of $G \setminus H$ is of this form.

It remains to investigate, which representations F^* and F' resp. F'' have to run through such that F runs exactly through a complete system of pairwise inequivalent and irreducible K-representations of $G \setminus H$.

Using Clifford's notation, we call two irreducible representations of $G \setminus H$ associated (with respect to G^*), if their restrictions to G^* have an irreducible constituent in common. From Clifford's theory we know, that there is a 1-1-correspondence between the classes of associated representations of $G \setminus H$ and the classes of representations of G^* which are conjugates with respect to $G \setminus H$. Two representations F^* and F^{**} of G^* are conjugates with respect to $G \setminus H$, if there is an $(f;\pi) \in G \setminus H$ so that

5.19 $$F^{*(f;\pi)} \sim F^{**} .$$

And this correspondence is as follows: the restriction to G^* of every element out of a class of associated representations is (up to its multiplicity) just the corresponding class of conjugate representations.

Hence it suffices, that in F the representation F^* runs through a complete system of representatives of the classes of conjugate representations of G*. Moreover Clifford's theory yields, that associated representations differ only in the factor, which is an irreducible representation of the inertia factor. Hence it suffices, that - while F* is fixed - F' runs through a complete system of irreducible K-representations of H'_{F^*}. Thus we have obtained the following theorem:

5.20 The irreducible K-representation $F = (\widetilde{F}^* \otimes F') \uparrow G \sim H$ runs exactly through a complete system of pairwise inequivalent and irreducible K-representations of $G \sim H$ if F* runs through a complete system of pairwise not conjugate but irreducible K-representations of G*, and, while F* is fixed, F" runs through a complete system of pairwise inequivalent K-representations of $H \cap S_{(n)}$.

In the special case $H = S_n$ two representations of G* are conjugates if and only if they are of the same type. Hence F runs through a complete system of irreducible K-representations of $G \sim S_n$ if F* runs through a complete system of irreducible K-representations with pairwise different types and F" - while F* is fixed - runs through a complete system of pairwise inequivalent and irreducible K-representations of $S_{(n)}$.

Thus the number of irreducible ordinary representations of $G \sim S_n$ is

<u>5.21</u>
$$\sum_{(n)} p(n_1)\ldots p(n_s) ,$$

if s is the number of conjugacy classes of G, p(m) is the number of partitions of m, p(0) := 1, and if the sum is taken over all the types $(n) = (n_1,\ldots,n_s)$.

This agrees with 3.8, the number of conjugacy classes of $G\sim S_n$.

Of course 5.21 yields also the number of irreducible K-representations if charK = p does not divide $|G\sim S_n|$ as long as K is algebraically closed. If charK = p | $|G\sim S_n|$ we have for the number of irreducible K-representations of $G\sim S_n$:

<u>5.22</u>
$$\sum_{(n)} pr(n_1)\ldots pr(n_t) ,$$

if t is the number of p-regular classes of G, pr(m) is the number of <u>p-regular</u> <u>partitions</u> of m (i.e. p doesn't divide the elements of the partition), pr(0) := 1, and the sum is taken over all the types $(n) = (n_1,\ldots,n_t)$.

As an example we derive the irreducible C-representations of the normalizer of $S_3 \times S_3$ in S_6, which is a faithful permutation representation of $S_3\sim S_2$ (cf. section 3):

(i) <u>The representations of the basis group S_3^*, their types, inertia groups and inertia factors:</u>

$S_3^* \simeq S_3 \times S_3$ has the following irreducible C-representations:

$[3]\#[3]$ $[3]\#[2,1]$ $[3]\#[1^3]$ $[2,1]\#[3]$ $[2,1]\#[2,1]$

$[2,1]\#[1^3]\quad [1^3]\#[3]\quad [1^3]\#[2,1]\quad [1^3]\#[1^3]$.

With respect to the arrangement $[3]$, $[2,1]$, $[1^3]$ of the irreducible C-representations of S_3, the types of these representations are

(2,0,0) (1,1,0) (1,0,1) (1,1,0) (0,2,0)

(0,1,1) (1,0,1) (0,1,1) (0,0,2) .

Hence a complete system of irreducible C-representations of S_3^* with pairwise different types is

$\{[3]\#[3],[3]\#[2,1],[3]\#[1^3],[2,1]\#[2,1],[2,1]\#[1^3],[1^3]\#[1^3]\}$.

The corresponding inertia groups are:

$S_3 \sim S_2,\ S_3^*,\ S_3^*,\ S_3 \sim S_2,\ S_3^*,\ S_3 \sim S_2,$

the inertia factors:

$S_2',\ S_1',\ S_1',\ S_2',\ S_1',\ S_2'.$

(ii) <u>The representations of $S_3 \sim S_2$</u>:

The irreducible ordinary representations of S_2 are $[2]$ and $[1^2]$, the only one of S_1 is $[1]$. Thus we get for the irreducible C-representations of $S_3 \sim S_2$:

$$\widetilde{[3]\#[3]} \otimes [2]' = \widetilde{[3]\#[3]}\ ,$$

$$\widetilde{[3]\#[3]} \otimes [1^2]'\ ,$$

$$(\widetilde{[3]\#[2,1]} \otimes [1]') \uparrow S_3 \sim S_2 = [3]\#[2,1] \uparrow S_3 \sim S_2\ ,$$

$$(\widetilde{[3]\#[1^3]} \otimes [1]') \uparrow S_3 \sim S_2 = [3]\#[1^3] \uparrow S_3 \sim S_2\ ,$$

$$\widetilde{[2,1]\#[2,1]} \otimes [2]' = \widetilde{[2,1]\#[2,1]}\ ,$$

$$\widetilde{[2,1]\#[2,1]} \otimes [1^2]'\ ,$$

$$(\widetilde{[2,1]\#[1^3]} \otimes [1]') \uparrow S_3 \sim S_2 = [2,1]\#[1^3] \uparrow S_3 \sim S_2\ ,$$

$$\widetilde{[1^3]\#[1^3]} \otimes [2]' = \widetilde{[1^3]\#[1^3]} \; ,$$

$$\widetilde{[1^3]\#[1^3]} \otimes [1^2]' \; .$$

Their degrees are 1,1,4,2,4,4,4,1,1 in accordance with

$$1^2 + 1^2 + 4^2 + 2^2 + 4^2 + 4^2 + 4^2 + 1^2 + 1^2 = 72 = |S_3 \sim S_2| .$$

(iii) <u>Representing matrices:</u>

As a numerical example we shall evaluate the matrix of $\widetilde{[2,1]\#[2,1]} \otimes [1^2]'$ representing (14)(25)(36), the image of (e;(12)) under the permutation representation of $S_3 \sim S_2$. (To get all the elements of a representation it suffices to evaluate the matrices representing generating elements. For generators of wreath products $S_m \sim S_n$ see Neumann [1].)

First, using the results of section 4, we obtain the matrix of [2,1]#[2,1] representing (e;1):

$$[2,1]\#[2,1](e;1) = \begin{matrix} {\scriptstyle \begin{smallmatrix}12\\3\end{smallmatrix}} & {\scriptstyle \begin{smallmatrix}13\\2\end{smallmatrix}} \end{matrix} \begin{bmatrix} 1 & 0 \\ 0 & 1 \end{bmatrix} \times \begin{matrix} {\scriptstyle \begin{smallmatrix}45\\6\end{smallmatrix}} & {\scriptstyle \begin{smallmatrix}46\\5\end{smallmatrix}} \end{matrix} \begin{bmatrix} 1 & 0 \\ 0 & 1 \end{bmatrix}$$

$$= \begin{matrix} {\scriptstyle \begin{smallmatrix}12 & 45\\3 & 6\end{smallmatrix}} & {\scriptstyle \begin{smallmatrix}12 & 46\\3 & 5\end{smallmatrix}} & {\scriptstyle \begin{smallmatrix}13 & 45\\2 & 6\end{smallmatrix}} & {\scriptstyle \begin{smallmatrix}13 & 46\\2 & 5\end{smallmatrix}} \end{matrix}$$

$$= \begin{bmatrix} 1 & 0 & 0 & 0 \\ 0 & 1 & 0 & 0 \\ 0 & 0 & 1 & 0 \\ 0 & 0 & 0 & 1 \end{bmatrix}$$

$$= (f^1_{\alpha_1\beta_1}(1) f^2_{\alpha_2\beta_2}(1)) \; ,$$

$$\Rightarrow \widetilde{[2,1]\#[2,1]}(e;(12)) = (f^1_{\alpha_1\beta_2}(1)f^2_{\alpha_2\beta_1}(1)) = \begin{bmatrix}1&0&0&0\\0&0&1&0\\0&1&0&0\\0&0&0&1\end{bmatrix} .$$

(We have to permute the second and third column.)

On account of $[1^2](12) = (-1)$ we obtain therefrom:

$$\widetilde{[2,1]\#[2,1]} \otimes [1^2]' (e;(12)) = \begin{bmatrix}1&0&0&0\\0&0&1&0\\0&1&0&0\\0&0&0&1\end{bmatrix} \times [-1]$$

$$= \begin{bmatrix}-1&0&0&0\\0&0&-1&0\\0&-1&0&0\\0&0&0&-1\end{bmatrix} .$$

This is a detailed description of the example Robinson gave (Robinson [5], 3.515).

Having now given a complete and detailed description of the construction of the irreducible representations of $G\sim H$ over an algebraically closed field we may now turn to special cases.

The procedure becomes much simpler if G is an abelian group. Special cases of this, namely wreath products of the form $C_m\sim S_n$ resp. $C_m\sim A_n$ of cyclic groups with symmetric resp. alternating groups have been considered by Young, Robinson, Osima, Puttaswamaiah and Frame (see Young [1], Robinson [1], Osima [1],[3], Puttaswamaiah [1],[2], Frame [1]), whose results can now be generalized.

If G is abelian, F* is onedimensional such that

5.23 $$\widetilde{F}^*(f;\pi) = F^*(f;1_H) = \prod_i F_i(f(i)) .$$

If $\pi_1,\dots,\pi_{|H:H\cap S_{(n)}|}$ is a complete system of representations of the left cosets of the inertia factor $H\cap S_{(n)}$ of F^* in H and F'' is an irreducible representation of $H\cap S_{(n)}$ we denote (see the notation of Curtis/Reiner [1]):

$$(\tilde{F}^* \,\dot{\otimes}\, F')(f_{\pi_i^{-1}};\pi_i^{-1}\pi\pi_k) := \begin{cases} (\tilde{F}^* \otimes F')(f_{\pi_i^{-1}};\pi_i^{-1}\pi\pi_k), & \text{if } \pi_i^{-1}\pi\pi_k \in S_{(n)} \\ 0 & \text{elsewhere .} \end{cases}$$

With this notation and 5.23 we obtain, if G is abelian:

$$\begin{aligned} F(f;\pi) &= ((\tilde{F}^* \,\dot{\otimes}\, F')(f_{\pi_i^{-1}};\pi_i^{-1}\pi\pi_k)) \\ &= (F^*(f_{\pi_i^{-1}};1_H)\dot{F}''(\pi_i^{-1}\pi\pi_k)), \end{aligned}$$

if

$$\dot{F}''(\pi_i^{-1}\pi\pi_k) := \begin{cases} F''(\pi_i^{-1}\pi\pi_k), & \text{if } \pi_i^{-1}\pi\pi_k \in H\cap S_{(n)} \\ 0 & \text{, elsewhere .} \end{cases}$$

Since

$$(F'' \uparrow H)(\pi) = (\dot{F}''(\pi_i^{-1}\pi\pi_k))$$

we have obtained

5.24 <u>If G is abelian we have for the representing matrices of $F = (\tilde{F}^* \otimes F') \uparrow G \sim H$:</u>

$$F(f;\pi) = (F^*(f_{\pi_i^{-1}};1_H)\cdot\dot{F}''(\pi_i^{-1}\pi\pi_k)) ,$$

<u>i.e. that the $|H:S_{(n)}|^2$ submatrices of which this matrix consists are up to the numerical factors $F^*(f_{\pi_i^{-1}};1_H)$</u>

equal to the submatrices of which the matrix $(F'' \uparrow H)(\pi)$ consists.

The evaluation of the matrices of $F'' \uparrow H$ has been described for the special case $H = S_n$ in 4.49 in case that there are only two factors, what can easily be generalized.
This method describing the construction of the matrices of the irreducible representations of $G \sim H$ for any abelian G generalizes the results of Puttaswamaiah for $C_m \sim S_n$ (Puttaswamaiah [2]) to wreath products $G \sim S_n$, G abelian, and the results of Frame on $C_2 \sim S_n$ (Frame [1]).

Let us return to 5.18. This theorem describes, how we can get the irreducible K-representations of $G \sim H$ from the irreducible K-representations of G and the irreducible K-representations of certain subgroups $H \cap S_{(n)}$ of H. Thus the representation theory of $S_m \sim S_n$ can be derived to a large extent with the aid of the representation theory of the symmetric group. This will be shown below, where we shall consider especially the repercussion of this fact on the representation theory of the symmetric group, for example on the theory of the symmetrized outer products $[\alpha] \odot [\beta]$ of irreducible ordinary representations $[\alpha]$ and $[\beta]$ of symmetric groups.
To generalize this theory of symmetrized outer products of symmetric groups we point now to certain irreducible representations

of $G \sim H$ with the aid of which we shall define symmetrized outer products for any representations of any permutation groups such that $[\alpha]\odot[\beta]$ is a special case.

At first we indicate certain irreducible K-representations of $G \sim H$ by a special notation: If F^* is of type $(0,\ldots,0,n,0,\ldots,0)$, i.e. if all the factors F_i of $F^* = \# F_i$ are equivalent to a certain irreducible K-representation of G, say to F^j, then we denote this by

<u>5.25</u> $$(F^j;F") := \widetilde{F^j \# \ldots \# F^j} \otimes F'$$

(The inertia factor of $F^* = \overset{n}{\#} F^j$ is $H' \cap S_n' = H'$, hence $F"$ is an irreducible K-representation of H.)

The special case of the representations $(\alpha;\beta)$ of a subgroup $S_m \sim S_n$ of S_{mn} (cf. 2.33) plays an important role in the ordinary representation theory of the symmetric group. The induced representations $(\alpha;\beta) \uparrow S_{mn} =: [\alpha]\odot[\beta]$ are the so-called symmetrized outer products of irreducible C-representations of symmetric groups (cf. section 6). Thus we get in a natural way the following generalization of this concept (Kerber [4]):

<u>5.26</u> Let $G \leq S_m$, $H \leq S_n$. Then we can identify $G \sim H$ with a subgroup of S_{mn} (cf. 2.24/2.25), and if F_G and F_H are any two K-representations of G and H (for any groundfield K), then we call

$$F_G \odot F_H := (F_G;F_H) \uparrow S_{mn} = (\widetilde{\overset{n}{\#} F_G}) \otimes F_H' \uparrow S_{mn}$$

<u>the symmetrized outer product of</u> F_G <u>and</u> F_H .

A rule which is obvious from the foregoing considerations is: If

$$F_H \leftrightarrow \sum_k a_k \hat{F}^k$$

indicates the decomposition of F_H into irreducible composition factors, then obviously $F_G \odot F_H$ has the same decomposition as

$$\sum_k a_k (F_G \odot \hat{F}^k) \ .$$

Hence in case of complete reducibility the symmetrized outer product multiplication is additive on the right hand side:

5.27 <u>If H is completely reducible over K, then</u>

$$F_G \odot \sum_k a_k \hat{F}^k = \sum_k a_k (F_G \odot \hat{F}^k) \ .$$

This generalizes a well known rule for the symmetrized outer product of C-representations of symmetric groups.

We would like to consider the characters of $G \sim H$ for a moment to derive some of their properties which will be of use later on.

If the factor F^* out of $F = (\tilde{F}^* \otimes F') \uparrow G \sim H$ is of type $(n) = (n_1, \ldots, n_r)$, then obviously every element $(f;\pi)$ whose permutation π has a partition $P\pi$ which is not a subpartition of the type (n) has zero as character value under F:

5.28 <u>If the factor F^* out of $F = (\tilde{F}^* \otimes F') \uparrow G \sim H$ is of type $(n) = (n_1, \ldots, n_r)$, then all the elements $(f;\pi)$ have 0 as character value under F whose conjugacy class has an empty intersection with $S_{(n)}$.</u>

For under this assumption not any one of the conjugates

$$(e;\pi_i)^{-1}(f;\pi)(e;\pi_i) = (f_{\pi_i^{-1}};\pi_i^{-1}\pi\pi_i)$$

of $(f;\pi)$ is contained in the inertia group from which F is induced, since for every i: $\pi_i^{-1}\pi\pi_i \notin H\cap S_{(n)}$. Thus we have only 0-matrices along the leading diagonal of $F(f;\pi)$.

q.e.d.

The difficulty of getting the characters of wreath products explicitly, except for special cases, arises from the fact that to pass from $F^*(f;1_H)$ to $F^*(f;\pi)$ we have to permute the columns of $F^*(f;1_H)$ so that the leading diagonal will be disturbed. For a discussion of the evaluation of these characters the reader is referred to Littlewood [3]. A useful remark is:

<u>5.29</u> <u>If π is an n-cycle then the character of $(f;\pi)$ vanishes under all the irreducible representations of $G\wr H$ which are not of the form $(F^j;F'')$.</u>

<u>Under $(F^j;F'')$ the element $(e;\pi)$ has the character value</u>

$$\zeta^{(F^j;F'')}(e;\pi) = f^{F^j}\zeta^{F''}(\pi) ,$$

<u>if f^{F^j} is the dimension of F^j.</u>

Proof: The first statement follows from 5.28.

Furthermore we know that

$$(F^j;F'')(e;\pi) = \widetilde{E_{(f^{F^j})^n}} \times F''(\pi) .$$

And in this first factor on the right hand side, which arises from the $(f^{F^j})^n$-rowed identity matrix by certain columnpermutations the leading diagonal contains, except 0's, as many 1's as is the dimension of F^j, for exactly the elements $f_{ii}^{F^j} \ldots f_{ii}^{F^j}$ $(1 \leq i \leq f^{F^j})$ remain in this leading diagonal (cf. 5.13).

q.e.d.

We would like now to derive some results concerning the modular representation theory of wreath products (cf. Kerber [2]) which will be of use later on in applying this theory to the modular representation theory of the symmetric group.

First we consider the case where G is a p-group. Then $G \sim H$ contains with its basis group a normal divisor which is a p-group as well as its centralizer if $G \neq \{1\}$.

For if $(f;1_H) \in G_i$, $f \neq e$, $\pi \neq 1$, we have

$$1 \neq (e;\pi)(f;1_H)(e;\pi)^{-1} = (f_\pi;1_H) \in G_{\pi^{-1}(i)} \neq G_i \,.$$

Hence the following is valid:

5.30 $$G \neq \{1\} \Rightarrow C_{G \sim H}(G^*) \leq G^*.$$

Thus we can apply a lemma of Brauer (Brauer [1], lemma 2) which says, that a group possesses exactly one p-block if it contains a normal p-group which includes its own centralizer. We have obtained:

5.31 If $G \neq \{1\}$ is a p-group, then $G \sim H$ possesses exactly one p-block.

Furthermore in this case the only irreducible p-modular representation of G* is the identity representation so that each p-modular irreducible representation of G∖H is of the form

5.32 $$\widetilde{IG^*} \otimes F' = F' .$$

Since $G \sim H = G^*H'$, the p-regular elements are contained in H' so that an ordinary irreducible representation of G∖H has the same Brauer character as its restriction to H':

5.33 <u>If G is a p-group, then a p-modular representation associated with an ordinary irreducible representation F of G∖H has the same decomposition numbers as $F \downarrow H'$.</u>

In case that G is an abelian p-group, the factor F* is onedimensional so that

$$\widetilde{F}^*(f;\pi) = F^*(f;1_H)$$

and the associated p-modular representation is the identity representation. Hence the following is valid:

5.34 <u>If G is an abelian p-group, then a p-modular representation associated with the ordinary irreducible representation $F = (\widetilde{F}^* \otimes F') \uparrow G\sim H$ has the same decomposition numbers as a p-modular representation associated with $F'' \uparrow H$.</u>

As an example we would like to evaluate the decomposition matrix of $S_2 \sim S_3$ with respect to p=2.

Since S_3 possesses two 2-blocks (see section 7 or the general representation theory of finite groups) and $\overline{[3]} = \overline{[1^3]}$, S_3 has as decomposition matrix with respect to p = 2:

$$\begin{bmatrix} 1 & 0 \\ 1 & 0 \\ 0 & 1 \end{bmatrix} \begin{matrix} [3] \\ [1^3] \\ [2,1] \end{matrix} \quad .$$

$S_2 \wr S_3$ has the ordinary irreducible representations

$$(2;3)\ , \qquad (2;2,1)\ , \qquad (2;1^3)\ ,$$
$$(\widetilde{[2]\#[2]\#[1^2]} \otimes ([2]\#[1])') \uparrow S_2\wr S_3\ ,$$
$$(\widetilde{[2]\#[2]\#[1^2]} \otimes ([1^2]\#[1])') \uparrow S_2\wr S_3\ ,$$
$$(\widetilde{[2]\#[1^2]\#[1^2]} \otimes ([1]\#[2])') \uparrow S_2\wr S_3\ ,$$
$$(\widetilde{[2]\#[1^2]\#[1^2]} \otimes ([1]\#[1^2])') \uparrow S_2\wr S_3\ ,$$
$$(1^2;3)\ , \qquad (1^2;2,1)\ , \qquad (1^2;1^3)\ .$$

Hence from 5.34 we obtain for the decomposition matrix (using the branching theorem to get $[2]\#[1] \uparrow S_3 = [2] \uparrow S_3 = [3] + [2,1]$, $[1^2]\#[1] \uparrow S_3 = [1^2] \uparrow S_3 = [2,1] + [1^3]$):

5.35
$$\begin{bmatrix} 1 & 0 \\ 0 & 1 \\ 1 & 0 \\ 1 & 1 \\ 1 & 1 \\ 1 & 1 \\ 1 & 1 \\ 1 & 0 \\ 0 & 1 \\ 1 & 0 \end{bmatrix} \begin{matrix} (2;3) \\ (2;2,1) \\ (2;1^3) \\ (\widetilde{[2]\#[2]\#[1^2]} \otimes ([2]\#[1])') \uparrow S_2\wr S_3 \\ (\widetilde{[2]\#[2]\#[1^2]} \otimes ([1^2]\#[1])') \uparrow S_2\wr S_3 \\ (\widetilde{[2]\#[1^2]\#[1^2]} \otimes ([1]\#[2])') \uparrow S_2\wr S_3 \\ (\widetilde{[2]\#[1^2]\#[1^2]} \otimes ([1]\#[1^2])') \uparrow S_2\wr S_3 \\ (1^2;3) \\ (1^2;2,1) \\ (1^2;1^3) \end{matrix}$$

If on the other hand the order $|G|$ of G is relatively prime to p, the situation is quite different. For then every p-modular representation associated to $\tilde{F}^*$ is irreducible since $\overline{\tilde{F}^* \downarrow G^*} = \overline{F^*}$ is irreducible and we obtain with a lemma of Osima (Osima [3], lemma 5):

5.36 <u>If $(|G|,p) = 1$ the p-block of $G\wr H$ to which $F=(\tilde{F}^* \otimes F')\uparrow G\wr H$ belongs has the same decomposition matrix as the block of the inertia factor to which F'' belongs.</u>
<u>$(\tilde{F}^* \otimes F_1') \uparrow G\wr H$ and $(\tilde{F}^* \otimes F_2') \uparrow G\wr H$ belong to the same block of $G\wr H$ if and only if F_1' and F_2' belong to the same block of the inertia factor $H' \cap S_{(n)}'$ of F^*.</u>

Using this and the well known decomposition matrix

$$\begin{bmatrix} 1 & 0 \\ 1 & 1 \\ 0 & 1 \end{bmatrix} \begin{matrix} [3] \\ [2,1] \\ [1^3] \end{matrix}$$

of S_3 with respect to $p = 3$ we obtain as decomposition matrix of $S_2\wr S_3$ for $p = 3$ and the same arrangement as in 5.35 of the rows:

5.37
$$\begin{bmatrix} 1 & 0 & & & & & & \\ 1 & 1 & & & & & & \\ 0 & 1 & & & & & & \\ & & 1 & & & & & \\ & & & 1 & & & & \\ & & & & 1 & & & \\ & & & & & 1 & & \\ & & & & & & 1 & 0 \\ & & & & & & 1 & 1 \\ & & & & & & 0 & 1 \end{bmatrix}.$$

It is more difficult to evaluate the decomposition matrix if neither G is an abelian p-group nor $(|G|,p) = 1$. But since 5.18 is independent of the characteristic we can sometimes proceed using this fact. As an example we evaluate the decomposition matrix of $S_3 \sim S_2$ for $p = 2$.

$\overline{[3]}$ and $\overline{[2,1]}$ are the 2-modular irreducible representations of S_3, $\overline{[2]}$ is the only one of S_2. Hence

$$\widetilde{\overline{[3]}\#\overline{[3]}} \otimes \overline{[2]}' ,$$

$$(\widetilde{\overline{[3]}\#\overline{[2,1]}} \otimes (\overline{[1]}\#\overline{[1]})') \uparrow S_3 \sim S_2 = \overline{[3]}\#\overline{[2,1]} \uparrow S_3 \sim S_2 ,$$

$$\widetilde{\overline{[2,1]}\#\overline{[2,1]}} \otimes \overline{[2]}'$$

are the 2-modular irreducible representations of $S_3 \sim S_2$.

The table of Brauer characters of this group is therefore

	$(1,1;1)$	$((123),1;1)$	$((123),(123);1)$
$\widetilde{\overline{[3]}\#\overline{[3]}} \otimes \overline{[2]}'$	1	1	1
$\widetilde{\overline{[2,1]}\#\overline{[2,1]}} \otimes \overline{[2]}'$	4	-2	1
$\overline{[3]}\#\overline{[2,1]} \uparrow S_3 \sim S_2$	4	1	-2

From this it follows that

5.38
$$\begin{array}{l} (3;2) \\ (3;1^2) \\ (1^3;1^2) \\ (1^3;2) \\ (2,1;2) \\ (2,1;1^2) \\ [3]\#[2,1] \uparrow S_3 \sim S_2 \\ [3]\#[1^3] \uparrow S_3 \sim S_2 \\ [2,1]\#[1^3] \uparrow S_3 \sim S_2 \end{array} \quad \begin{bmatrix} 1 & 0 & 0 \\ 1 & 0 & 0 \\ 1 & 0 & 0 \\ 1 & 0 & 0 \\ 0 & 1 & 0 \\ 0 & 1 & 0 \\ 0 & 0 & 1 \\ 2 & 0 & 0 \\ 0 & 0 & 1 \end{bmatrix}$$

is the decomposition matrix of $S_3 \sim S_2$ for $p = 2$ (cf. also the character table of $S_3 \sim S_2$ in section 6).

Concluding this section we point once more to the construction of the irreducible representations of $G \sim H$.

If we know how to construct the irreducible representations of G and of the subgroups $H \cap S_{(n)}$ of H we get the representing matrices $F(f;\pi)$ as we have described above: First we form Kronecker products of irreducible K-representations of G (see 5.1/5.2), then we have to permute the columns of these matrices (see 5.13). After this we have to form Kronecker products once more (see 5.17) and at last we have to induce (see 5.18) to get $F(f;\pi)$.

Hence the entries of $F(f;\pi)$ have all the properties which are properties of the entries of the irreducible representations of G and $H \cap S_{(n)}$ and which are invariant with respect to column permutation and the Kronecker product multiplication as well as the inducing process.

From this consideration we get some results about splitting fields (cf. Kerber [7]):

<u>5.39</u> <u>If $S \subset K$ (K algebraically closed) is a splitting field for G and the subgroups $H \cap S_{(n)}$, then S is a splitting field for $G \sim H$, too, i.e. every irreducible K-representation of $G \sim H$ is realizable in S.</u>

E. g. if all the ordinary irreducible representations of G and of the subgroups $H \cap S_{(n)}$ are realizable in R (resp. Q), then all the ordinary irreduzible representation of $G \sim H$ are realizable in R (resp. Q) as well. Thus for example all the ordinary irreducible representations of $S_m \sim S_n$ are realizable in Q and hence even in Z. A weaker assumption would be that the ordinary characters of G as well as of the subgroups $H \cap S_{(n)}$ are real. But it is possible, that this is not carried over to $G \sim H$ since the leading diagonal is disturbed when passing from $F^*(f;1_H)$ to $\tilde{F}^*(f;\pi)$. We know that the reality of the ordinary characters is equivalent to the ambivalency of the group. And as we have proved in section 3 (cf. 3.14) $G \sim S_n$ is ambivalent if G is ambivalent. Thus we have, though this does not follow from the construction of the matrices (Kerber [7]),:

5.40 <u>If the characters of G are real, the characters of $G \sim S_n$ are real as well.</u>

From 3.16 we get:

5.41 <u>If the (ordinary) character table of G or of H is not real, then the character table of $G \sim H$ is complex.</u>

A finite group is called an <u>M-group</u>, if every irreducible representation over an algebraically closed field whose characteristic doesn't divide $|G|$ is induced by a onedimensional representation of a suitable subgroup. Another corollary to 5.18 concerning this conception is (Seitz [1], Kerber [6]):

__5.42__ __If G and the subgroups $H \cap S_{(n)}$ are M-groups, then $G \sim H$ is an M-group.__

Proof: If K is algebraically closed and $\operatorname{char} K \nmid |G|$, then every irreducible K-representation of G and hence also of G^* is equivalent to a representation, whose matrices contain in every row and in every column exactly one nonvanishing entry. This is valid for F^* as well as for F'' and this is a property invariant under column permutation, Kronecker product multiplication and the inducing process. Hence it is a property of $\tilde{F}^*$ and F'' and $\tilde{F}^* \otimes F'$ and $F = (\tilde{F}^* \otimes F') \uparrow G \sim H$. From a well known theorem (see Huppert [1], V, 18.9) it follows, that since F is irreducible, F is induced by a onedimensional representation of a suitable subgroup.

q.e.d.

E.C. Dade has proved this for the special case $H = C_p = \langle(1 \ldots p)\rangle \leq S_p$ (p a prime number, see Huppert [1], V, 18.10) as an important part of his proof, that every solvable group can be imbedded in an M-group (see Huppert [1], V, 18.11).

The last 4 theorems are corollaries of the construction of the irreducible representations of $G \sim H$ we have given above. Another theorem concerning generalized decomposition numbers will be given in section 8.

With the aid of the results of this section we would like now to describe the application of this theory of wreath products to the representation theory of symmetric and alternating groups.

Chapter III

Application to the representation theory

of symmetric and alternating groups

Two examples, the theory of the symmetrized outer products and the theory of the generalized decomposition numbers will show how the representation theory of wreath products can be applied to the representation theory of symmetric and alternating groups.
More precisely we apply the representation theory of wreath products of the form $G \sim S_n$. In the first case the used subgroups are of the form $S_m \sim S_n$, in the second case of the form $C_m \sim S_n$.
Although the representation theory of $G \sim S_n$ has been derived with the aid of the representation theory of the symmetric group, our argument is not circular since for the applied subgroups $G \sim S_m \leq S_n$ we have $m<n$. Hence this application is actually a recursion pro-

cess. Thus for example the evaluation of the (strictly) generalized decomposition numbers of S_n is reduced to the evaluation of decomposition numbers of symmetric groups of lower degrees $m<n$.

There are probably other ways of applying this theory of representations of wreath products to the theory of the symmetric group. Presumably we can illuminate in this way the concept of the so-called "star-diagram" or "p-quotient" of a Young-diagram satisfactorily, but this will be discussed in the following parts of this paper.

The first section of this chapter contains the theory of the symmetrized outer products of irreducible ordinary representations of symmetric groups. Then as a preparation for the third section which contains the theory of the generalized decomposition matrix of symmetric and alternating groups we summarize in the second section some known and some new results about decomposition numbers of symmetric and alternating groups.

6. Symmetrized outer products of irreducible C-representations of symmetric groups

A chance to apply the representation theory of wreath products to the ordinary representation theory of the symmetric group arises from the trivial fact, that the normalizer $N_{S_{mn}}(\overset{n}{\times} S_m)$ of a direct product $\overset{n}{\times} S_m := S_m \times \ldots \times S_m$ (n factors) of n subgroups isomorphic to S_m and in S_{mn} lies between $\overset{n}{\times} S_m$ and S_{mn}:

6.1 $$\overset{n}{\times} S_m \leq N_{S_{mn}}(\overset{n}{\times} S_m) \leq S_{mn} .$$

For as we have seen in section 2 (cf. 2.33) this normalizer is a faithful permutation representation of $S_m \sim S_n$ with $\overset{n}{\times} S_m$ as its basis group.

Hence the problem to derive the reduction of the representation $[\alpha]\ldots[\alpha] = [\alpha]\#\ldots\#[\alpha] \uparrow S_{mn}$ induced from $\overset{n}{\times} S_m$ can be divided into two problems. For using the transitivity of the inducing process and 6.1 we obtain:

6.2 $$[\alpha]\ldots[\alpha] = [\alpha]\#\ldots\#[\alpha] \uparrow N_{S_{mn}}(\overset{n}{\times} S_m) \uparrow S_{mn} .$$

The first one of the two remaining problems is to derive the reduction of

6.3 $$[\alpha]\#\ldots\#[\alpha] \uparrow N_{S_{mn}}(\overset{n}{\times} S_m) .$$

The second one is the problem to give the reduction of the representations of S_{mn} induced by the irreducible constituents of 6.3. Let us first consider the reduction of 6.3.

We know that $N_{S_{mn}}(\overset{n}{\times} S_m)$ is a faithful permutation representation of $S_m \sim S_n$, hence the problem is to give the reduction of

6.4 $$\overset{n}{\#}[\alpha] \uparrow S_m \sim S_n \ .$$

But with the reciprocity theorem of Frobenius and 5.18 we have at once the solution of this problem:

<u>**6.5**</u> $$\overset{n}{\#}[\alpha] \uparrow S_m \sim S_n = \sum_{\beta} f^{\beta}(\alpha;\beta),$$

if the sum is taken over all the partitions of n and if f^{β} denotes the dimension of $[\beta]$.

Applied to our starting problem we obtain from 6.5:

6.6 $$\overset{n}{\#}[\alpha] \uparrow S_{mn} = \sum_{\beta} f^{\beta}((\alpha;\beta) \uparrow S_{mn}) \ ,$$

Thus we have reduced this problem to the reduction of the <u>symmetrized</u> <u>outer</u> <u>products</u> (cf. 5.26)

6.7 $$[\alpha]\odot[\beta] := (\alpha;\beta) \uparrow S_{mn} \ .$$

Gathering up we obtain the equation

<u>**6.8**</u> $$[\alpha]\ldots[\alpha] = \sum_{\beta} f^{\beta}([\alpha]\odot[\beta]) \ ,$$

which originally directed the attention to certain in general reducible representations of S_{mn} which were called symmetrized outer products. A hint to clarify their theory with the aid of the representation theory of wreath products we owe to Robinson who pointed out that they are induced by certain irreducible representations of $N_{S_{mn}}(\overset{n}{\times} S_m)$ (see Robinson [3],[4],[5], Kerber [4]).

To derive results of the theory of symmetrized outer products it is useful to describe the ordinary representation theory of $S_m \sim S_n$ in more detail. This we shall do now using the results of the sections 4 and 5. But we shall not only consider the irreducible representations of the form $(\alpha;\beta)$.

From the results of the sections 4 and 5 we obtain, that all the irreducible $\mathbb{C}$-representations of $S_m \sim S_n$ are of the form

<u>6.9</u> $$F = ((\widetilde{\overset{n}{\underset{j=1}{\#}} [\alpha]_j}) \otimes (\overset{p(m)}{\underset{k=1}{\#}} [\beta]_k)') \uparrow S_m \sim S_n$$

with irreducible $\mathbb{C}$-representations $[\alpha]_j$ of S_m and $[\beta]_k$ of S_{n_k}, if $\#[\alpha]_j$ is of type $(n_1,\ldots,n_{p(m)})$.

The number of irreducible $\mathbb{C}$-representations of $S_m \sim S_n$ is (cf.5.21):

<u>6.10</u> $$\sum_{(n)} p(n_1)\ldots p(n_{p(m)}) ,$$

if the sum is taken over all the p(m)-tupels $(n) = (n_1,\ldots n_{p(m)})$ so that $0 \leq n_i \in \mathbb{Z}$, $\Sigma\, n_i = n$.

We have given an example for the evaluation of the representing matrices in the last section.

p(m)p(n) of these representations 6.9 are of the special form $(\alpha;\beta)$.

The next question is, how we can evaluate the character table of

$S_m \sim S_n$. Of course we can obtain the characters by evaluating the matrices for a complete set of representatives of the conjugacy classes under each irreducible ordinary representation and checking the traces of these matrices. But this is the most complicated way to get the character table and it can be simplified very much by evaluating the matrices only for certain representations and using symmetries of these tables.

These symmetries are of the same kind as these of the character table of S_n which arise from the fact described by 4.11 from what follows that we get the row of $[\alpha']$ from the row of $[\alpha]$ by changing the sign in the columns of conjugacy classes which belong to $S_n \backslash A_n$.

The reason for these symmetries of the character table of S_n is the fact that $[\alpha]$ and $[\alpha']$ form a pair of irreducible $\mathbb{C}$-representations of S_n which are associated (in the sense of Clifford's theory of representations of groups with normal subgroups) with respect to the normal divisor A_n of index 2.

This can be applied also to $S_m \sim S_n$, even with more success: The normalizer $N_{S_{mn}}(\overset{n}{\times} S_m)$, a faithful permutation representation of $S_m \sim S_n$, contains for m,n>1 two different and nontrivial normal subgroups of index 2, the subgroup

6.11 $$S_m \sim S_n^{+} := S_m \sim S_n \cap A_{mn}$$

of the even permutations, and the subgroup

6.12 $S_m \sim A_n$.

(For the sake of simplicity we write $S_m \sim S_n{}^+$ instead of $(S_m \sim S_n)^+$, hoping that this subgroup will not be confused with $S_m \sim S_n^+ = S_m \sim A_n$.) Let us denote by F^+ the representation associated with F with respect to $S_m \sim S_n{}^+$ and by F^A the representation associated with F with respect to $S_m \sim A_n$. Thus F^+ (resp. F^A) means the inner tensor product of F and the alternating representation of $S_m \sim S_n$ with respect to $S_m \sim S_n{}^+$ (resp. $S_m \sim A_n$). If we denote these alternating representations by $A_+ S_m \sim S_n$ resp. $A_A S_m \sim S_n$ we have obviously

6.13
$$A_+ S_m \sim S_n = \begin{cases} (1^m; n), \text{ if } m \text{ is even} \\ (1^m; 1^n), \text{if } m \text{ is odd} \end{cases} ,$$

$$A_A S_m \sim S_n = (m; 1^n) .$$

Thus

6.14
$$F^+ = \begin{cases} F \otimes (1^m; n) , \text{ if } 2 \mid m \\ F \otimes (1^m; 1^n), \text{ if } 2 \nmid m , \end{cases}$$

$$F^A = F \otimes (m; 1^n) .$$

To describe F^+ and F^A explicitly in the form 6.9 we take into account that for inner tensor products the following is valid if $H \leq G$, D_1 is a representation of H, D_2 a representation of G:

$$(D_1 \uparrow G) \otimes D_2 = (D_1 \otimes (D_2 \downarrow H)) \uparrow G .$$

Using this it is easy to verify the following theorem (Kerber [4]):

6.15 <u>The irreducible C-representations of $S_m \sim S_n$ associated with the representation 6.9 with respect to $S_m \sim S_n{}^+$ resp. $S_m \sim A_n$ are:</u>

$$F^+ := \begin{cases} ((\widetilde{\#[\alpha']_j}) \otimes (\#[\beta]_k)') \uparrow S_m \wr S_n \ , & \text{if } 2 \mid m \\ ((\widetilde{\#[\alpha']_j}) \otimes (\#[\beta']_k)') \uparrow S_m \wr S_n, & \text{if } 2 \nmid m \end{cases}$$

resp.

$$F^A := ((\widetilde{\#[\alpha]_j}) \otimes (\#[\beta']_k)') \uparrow S_m \wr S_n \ ,$$

if $[\alpha']_j$ resp. $[\beta']_k$ denotes the representation associated to $[\alpha]_j$ resp. $[\beta]_k$ with respect to A_m resp. A_{n_k}, i.e. if

$$[\alpha']_j := [\alpha]_j \otimes [1^m] \ , \ [\beta']_k := [\beta]_k \otimes [1^{n_k}] \ .$$

Special cases are

$$(\alpha;\beta)^+ = \begin{cases} (\alpha';\beta) \ , & \text{if } 2 \mid m \\ (\alpha';\beta'), & \text{if } 2 \nmid m \ , \end{cases}$$

$$(\alpha;\beta)^A = (\alpha;\beta') \ .$$

This indicates how we can get the rows of F^+ and F^A from the row of F by changing the sign in certain columns.

F is called selfassociated with respect to $S_m \wr S_n^+$ ($S_m \wr A_n$) if $F = F^+$ ($F = F^A$). The character of such a selfassociated representation vanishes outside of $S_m \wr S_n^+$ ($S_m \wr A_n$).

Finally we mention special character values which can be obtained by specializing 5.28 and 5.29:

6.16 If in 6.9 the representation $\#[\alpha]_j$ of the basis group is of type $(n_1,\ldots,n_{p(m)})$ and if the conjugacy class of π in S_n has an empty intersection with $S_{(n)}$, then we have for the

<u>ζ^F of F:</u>

$$\zeta^F(f;\pi) = 0, \ \forall f.$$

<u>If π is an n-cycle, then $\zeta^F(f;\pi) = 0$ if F is not of the form $(\alpha;\beta)$, and</u>

$$\zeta^{\alpha;\beta}(e;\pi) = \begin{cases} (-1)^k f^\alpha, & \text{if } [\beta] = [n-k,1^k] \ (0 \leq k < n) \\ 0 & \text{elsewhere} \end{cases}.$$

(For the last statement cf. 4.42)

As an example we give the character table of $S_3 \wr S_2$. The distribution of the elements of $S_3 \wr S_2$ into conjugacy classes we gave in section 3. The ordinary irreducible representations of this group have been described in section 5, it remains to evaluate the characters.

The first row of this table on page 123 is the trivial row of the identity representation $(3;2) = \widetilde{[3]\#[3]} \otimes [2]'$. The second and third row can be obtained from the first row according to 6.15 and the fourth row can be obtained from the second one.

We evaluate the matrices representing elements of the conjugacy classes under (2,1;1) and get therfrom the fifth row and from this the sixth row according to 6.15. Similarly we obtain the seventh row from which we get the ninth one. The eigth row belongs to a representation which is selfassociated with respect to $S_3 \wr S_2{}^+$ as well as $S_3 \wr A_2$ such that for this row there are only 3 matrices to

representatives of conjugacy classes:	(1,1;1)	((12),1;1)	((123),1;1)	((12),(12);1)	((123),(123);1)	((12),(123);1)	(1,1;(12))	((123),(123);(12))	((12),(123);(12))
" " " :	1	(12)	(123)	(12)(45)	(123)(456)	(12)(456)	(14)(25)(36)	(153426)	(15)(2634)
class orders:	1	6	4	9	4	12	6	12	18
$(3;2)$	1	1	1	1	1	1	1	1	1
$(3;1^2)=(3;2)^A$	1	1	1	1	1	1	-1	-1	-1
$(1^3;1^2)=(3;2)^+$	1	-1	1	1	1	-1	-1	-1	1
$(1^3;2)=(3;1^2)^+$	1	-1	1	1	1	-1	1	1	-1
$(2,1;2)$	4	0	-2	0	1	0	2	-1	0
$(2,1;1^2)=(2,1;2)^A$	4	0	-2	0	1	0	-2	1	0
$[3]\#[2,1]\uparrow S_3\sim S_2$	4	2	1	0	-2	-1	0	0	0
$[3]\#[1^3]\uparrow S_3\sim S_2$	2	0	2	-2	2	0	0	0	0
$[2,1]\#[1^3]\uparrow S_3\sim S_2$ $=([3]\#[2,1]\uparrow S_3\sim S_2)^+$	4	-2	1	0	-2	1	0	0	0

be constructed.

In the same way we get the zeros of the second, fourth and sixth column and the fifth and sixth row. The box containing zeros in the right and lower corner we get from 6.16 (cf. Kerber [4]). This table has also been evaluated by Littlewood (Littlewood [2], p. 275) with the aid of the table of S_6, while for this procedure to get the table of $S_3 \sim S_2$ (in the general case: $S_m \sim S_n$) we used only the representations of S_3 and S_2 (resp. S_m and S_n and subgroups of S_n in the general case) and not the table of S_6 (resp. S_{mn}). In Littlewood's book we can find also the tables of $S_4 \sim S_2$ (p. 277), $S_2 \sim S_4$ (p. 278) and $S_3 \sim S_3$ (p. 280). In a paper of Robinson (Robinson [4]) the table of $S_2 \sim S_3$ can be found, he also did not use the table of S_6. In a later paper Littlewood used this more direct method, too (Littlewood [3]).

Having derived these detailed results on the ordinary representation theory of $S_m \sim S_n$ we return to the theory of symmetrized outer products $[\alpha] \odot [\beta]$ of irreducible ordinary representations of symmetric groups.

A problem which is up to now only incompletely solved is the reduction of $[\alpha] \odot [\beta]$ (see the references given in Robinson [5] and Boerner [3]). But using our theorems we can easily obtain the most important results of this theory.

At first we get at once from 6.15 for the multiplicity

$([\alpha]\odot[\beta],[\gamma])$ of the irreducible representation $[\gamma]$ of S_{mn} in $[\alpha]\odot[\beta]$:

$$\underline{6.17} \qquad ([\alpha]\odot[\beta],[\gamma]) = \begin{cases} ([\alpha']\odot[\beta],[\gamma']) \ , \text{ if } 2 \mid m \\ ([\alpha']\odot[\beta'],[\gamma']), \text{ if } 2 \nmid m \ , \end{cases}$$

the so-called <u>Littlewood's theorem of conjugates</u> (Littlewood [1]). Another theorem, perhaps the most important one can also be obtained easily. With its help our problem can be reduced to the problem to give the reduction of symmetrized outer products of the special form $[\alpha]\odot[r]$ ($r\leq n$). The assertion is analogous to 4.41. If $n = n_1+n_2$ we denote at first

6.18 $\quad ([\alpha]\odot[n_1])([\alpha]\odot[n_2]) := (([\alpha]\odot[n_1])\#([\alpha]\odot[n_2])) \uparrow S_{mn}$

and prove that the following is valid:

$$\underline{6.19} \qquad ([\alpha]\odot[n_1])([\alpha]\odot[n_2]) = [\alpha]\odot([n_1][n_2]) \ .$$

Proof: Using a well known theorem about induced representations (see Curtis/Reiner [1],(43.2)) and the transitivity of the inducing process we obtain for the left hand side, using 6.18:

$$((\alpha;n_1)\#(\alpha;n_2)) \uparrow S_m\sim(S_{n_1}\times S_{n_2}) \uparrow S_m\sim S_n \uparrow S_{mn} \ .$$

Because of $(S_m\sim S_{n_1})\times(S_m\sim S_{n_2}) = S_m\sim(S_{n_1}\times S_{n_2})$ this is equal to

$$(\alpha;[n_1]\#[n_2]) \uparrow S_m\sim S_n \uparrow S_{mn} = [\alpha]\odot([n_1][n_2]) \ .$$

q.e.d.

Applying 4.41 to $[\beta]$:

$$[\beta] = |[\beta_i+j-i]|$$

we obtain the announced result (Robinson [5], 3.531):

6.20 $$[\alpha]\odot[\beta] = |[\alpha]\odot[\beta_i+j-i]| .$$

For example

$$[\alpha]\odot[2,1] = \begin{vmatrix} [\alpha]\odot[2] & [\alpha]\odot[3] \\ 1 & [\alpha]\odot[1] \end{vmatrix}$$

$$= ([\alpha]\odot[2])([\alpha]\odot[1]) - [\alpha]\odot[3]$$

$$= ([\alpha]\odot[2])[\alpha] - [\alpha]\odot[3] .$$

$[\alpha]\odot[\beta_i+j-i]$ is of the form $[\alpha]\odot[r]$, $r\leq n$. And if we know the reductions of these $[\alpha]\odot[\beta_i+j-i]$ we get the reduction of $[\alpha]\odot[\beta]$ with 6.20 and the Littlewood-Richardson-rule 4.51.

It is remarkable that with the reciprocity theorem of Frobenius we obtain

6.21 $$([\alpha]\odot[r],[\gamma]) = ([\gamma]\downarrow S_m \sim S_r, \widetilde{\overset{r}{\#}[\alpha]}) ,$$

which reduces the problem to the evaluation of the character of $\widetilde{\overset{r}{\#}[\alpha]}$ (cf. Kerber [4]).

Concluding this section we would like to derive two equations concerning the matrix

6.22 $$R^\alpha := (r^\alpha_{\gamma\beta})$$

of the multiplicities

6.23 $$r^\alpha_{\gamma\beta} := ([\alpha]\odot[\beta],[\gamma]) ,$$

for a fixed partition α of m and for the partitions β of n, γ of mn. Thus R^{α} is a p(mn)xp(n)-matrix.

Let $[\delta]$ be an irreducible representation of S_{n-1} and

6.24 $$r^{\alpha *}_{\varepsilon\delta} := ([\alpha]\odot[\delta],[\varepsilon]) .$$

Using the Littlewood-Richardson-rule 4.51 we get at once the matrices $S^{\alpha} := (s^{\alpha}_{\gamma\varepsilon})$ and $E := (e_{\beta\delta})$, whose entries are defined by

6.25 $$s^{\alpha}_{\gamma\varepsilon} := ([\varepsilon][\alpha],[\gamma])$$

and

6.26 $$e_{\beta\delta} := ([\delta][1],[\beta]) .$$

A special case of 6.19 is

6.27 $$([\alpha]\odot[\delta])[\alpha] = [\alpha]\odot([\delta][1]) .$$

We can compare the multiplicities of the irreducible constituents on both sides of 6.27.

For the left hand side we obtain from 6.24 and 6.25:

6.28 $$([\alpha]\odot[\delta])[\alpha] = \sum_{\varepsilon} r^{\alpha *}_{\varepsilon\delta}\,[\alpha] = \sum_{\varepsilon,\gamma} s^{\alpha}_{\gamma\varepsilon} r^{\alpha *}_{\varepsilon\delta}[\gamma] .$$

And since 6.26 is valid the right hand side of 6.27 is

6.29 $$[\alpha]\odot([\delta][1]) = \sum_{\beta} e_{\beta\delta}[\alpha]\odot[\beta] = \sum_{\beta,\gamma} r^{\alpha}_{\gamma\beta} e_{\beta\delta}[\gamma] .$$

Comparing 6.28 and 6.29 we obtain:

<u>6.30</u> $$R^{\alpha}E = S^{\alpha}R^{\alpha *}$$

(Robinson [5], 3.548) as a necessary condition for the wanted matrix R^{α}.

A second condition for R^{α} can be derived using the theorem 6.16 concerning the characters of the irreducible representations of

$S_m \sim S_n$.

We consider the equation 6.23. Using the reciprocity theorem of Frobenius we obtain

6.31 $$[\gamma] \downarrow S_m \sim S_n = \sum_{\alpha,\beta} r^{\alpha}_{\gamma\beta}(\alpha;\beta) + \dots ,$$

if we gather under the summation sign exactly those irreducible constituents of $[\gamma] \downarrow S_m \sim S_n$, which are of the form $(\alpha;\beta)$. Since 6.16 it suffices to regard only these constituents of the form $(\alpha;\beta)$ if we take the character of $(e;(1\dots n))$ on both sides of this equation. We obtain

$$\zeta^{\gamma}_{(n^m)} = \sum_{\alpha,\beta} r^{\alpha}_{\gamma\beta}\zeta^{\alpha;\beta}(e;(1\dots n)).$$

(The partition of $(e;(1\dots n))$ is $P(e;(1\dots n)) = (n,\dots,n) = (n^m)$.).

Using 6.16 we get therefrom:

6.32 $$\zeta^{\gamma}_{(n^m)} = \sum_{\alpha,\beta:\beta=(n-k,1^k)} r^{\alpha}_{\gamma\beta} f^{\alpha}(-1)^k$$

as a second necessary condition for the coefficients $r^{\alpha}_{\gamma\beta}$ of the wanted matrix R^{α} (see Kerber [4]).

At the end of this section we give three of the most important rules for these symmetrized outer products. They are well known but now we can obtain these rules much more directly with the aid of the representation theory of wreath products.

A generalization of 6.19 is

6.33 $$([\alpha]\odot[\beta])([\alpha]\odot[\gamma]) = [\alpha]\odot([\beta][\gamma]) ,$$

the proof is quite analogous to the proof of 6.19.

A special case of 5.27 is

6.34 $$[\alpha]\odot([\beta] + [\gamma]) = [\alpha]\odot[\beta] + [\alpha]\odot[\gamma] ,$$

and from the associativity of the wreath product multiplication we obtain

6.35 $$([\alpha]\odot[\beta])\odot[\gamma] = [\alpha]\odot([\beta]\odot[\gamma]).$$

7. Block-structure and decomposition numbers of symmetric and alternating groups

The development of the modular representation theory of the symmetric group began in 1940 with the publication of the two parts of T. Nakayama's paper "On some modular properties of irreducible representations of symmetric groups" (Nakayama [1],[2]). The second part of this paper concludes with a conjecture about the p-block-structure of the symmetric group which has been proved first by Brauer and Robinson in 1947 (Brauer [1], Robinson [2]) and which is the foundation for all the following papers concerning this theory.

Beyond this fundamental theorem which is still called "Nakayama's conjecture" we know a lot of results but not the wanted general results (i.e. independent of n and p) about the decomposition numbers of S_n with respect to p. Furthermore we know how the evaluation of the generalized decomposition numbers can be reduced to the evaluation of decomposition numbers of symmetric groups of lower degrees and we know some analogous theorems about the alternating group, e.g. we know the block-structure of A_n and how we can get the decomposition matrix of A_n as far as the decomposition matrix of S_n is known.

We would like to describe some of these newer results beginning with the theorem about the block-structure of A_n. We shall describe the present situation of the theory of decomposition numbers of symmetric and alternating groups.

To formulate Nakayama's conjecture we need the following definition:

7.1 Def.: Let $[\alpha]$ be a Young-diagram and q a natural number. If we cancel successively parts of the rim which belong to hooks of length q, then a (uniquely determined) subdiagram $[\tilde{\alpha}]$ remains which we call the *q-core* of $[\alpha]$.

For example if $q = 3$:

$$\Rightarrow \quad \widetilde{[3^2,2,1]} = [0] .$$

If p is a prime number, then *the fundamental theorem of the modular representation theory of the symmetric group* reads as follows:

7.2 (*"Nakayama's conjecture"*)

The irreducible ordinary representations of S_n which form the p-block of S_n to which $[\alpha]$ belongs are exactly the representations $[\beta]$ of S_n with the same p-core as $[\alpha]$, i.e. the $[\beta]$ with

$$[\tilde{\beta}] = [\tilde{\alpha}] .$$

As has been said above: this is no longer a conjecture, since this theorem has been proved in 1947 by Brauer and Robinson!

To prove 7.2 the well known characterization of the p-blocks by the class multipliers (see Curtis/Reiner [1], 85,86) can be used: $[\alpha]$ and $[\beta]$ belong to the same p-block if and only if their characters ζ^α, ζ^β satisfy the following congruence modulo p:

7.3 $$\omega_a^\alpha := (|C^a|/f^\alpha)\zeta_a^\alpha \equiv (|C^a|/f^\beta)\zeta_a^\beta =: \omega_a^\beta \mod p,$$

on all the conjugacy classes C^a of S_n.

An important role in the proof is played by the defect of a block. It turns out, that the defect d^α of the block containing $[\alpha]$ and the number b^α of p-hooks which can be removed from $[\alpha]$ to yield $[\tilde{\alpha}]$ satisfy the equation

<u>7.4</u> $$d^\alpha = b^\alpha + e_p(b^\alpha!) ,$$

if as in section 1 $e_p(m)$ denotes the exponent of the maximal power of p which divides m. b^α is called the <u>p-weight</u> of $[\alpha]$.

Actually the following is valid (Brauer [1]):

<u>7.5</u> <u>The defect group of the block of $[\alpha]$ is isomorphic to a p p-Sylow-subgroup of $C_p \wr S_{b^\alpha} \leq S_{pb^\alpha}$, i.e. isomorphic to the wreath product $C_p \wr P_{b^\alpha}$, if P_{b^α} denotes a p-Sylow-subgroup of S_{b^α}.</u>

This provides an opportunity to apply the theory of representations of wreath products. Actually some parts of the known proofs of Nakayama's conjecture can be simplified in this way. But this remark may suffice here, a more detailed discussion is left to the later parts of this paper, since we shall have a similar application to the theory of generalized decomposition numbers which will be discussed in full detail.

Before doing this we would like to consider the alternating group. The theorem concerning alternating groups and analogous to Nakayama's conjecture is (Puttaswamaiah [1], Puttaswamaiah/Robinson [1], Kerber [3]):

7.6 (i) If $[\alpha] = [\widetilde{\alpha}]$, then every irreducible constituent of $[\alpha]\downarrow A_n$ forms its own p-block and each modular representation associated with such a constituent is irreducible.

(ii) If $[\alpha] \neq [\widetilde{\alpha}]$, then to the p-block of an irreducible constituent of $[\alpha] \downarrow A_n$ belong exactly the irreducible constituents of restrictions $[\beta] \downarrow A_n$ of such representations $[\beta]$ of S_n for which

$$[\widetilde{\beta}] = [\widetilde{\alpha}] \quad \text{or} \quad [\widetilde{\beta}] = [\widetilde{\alpha}'] .$$

Proof: We shall repeatedly use the theorem (cf. Curtis/Reiner [1], (86.3)) that an ordinary irreducible representation forms its own p-block and is modular irreducible if p is contained in its dimension as often as in the order of the group.

(i) We assume first, that $[\alpha] = [\tilde{\alpha}]$.

a) If $\alpha \neq \alpha'$, then (cf. 4.54) $[\alpha] \downarrow A_n$ is irreducible.

$[\alpha] = [\tilde{\alpha}]$ implies, that $[\alpha]$ contains no hook of length p, hence in this case (cf. 4.46):

$e_p(f^{[\alpha]\downarrow A_n}) = e_p(f^\alpha) = e_p(n!) \geq e_p(n!/2) = e_p(|A_n|)$, if $n>1$.

Using the theorem mentioned above, this part of the statement is proved for the case $[\tilde{\alpha}] = [\alpha] \neq [\alpha']$.

b) If on the contrary $\alpha = \alpha'$, then by 4.54 the restriction decomposes:

$$[\alpha] \downarrow A_n = [\alpha]^+ + [\alpha]^-,$$

and $[\alpha]^\pm$ are irreducible ordinary representations of dimension $f^\alpha/2$. $[\alpha] = [\tilde{\alpha}]$ implies again $e_p(n!) = e_p(f^\alpha)$. Thus

$$e_p(f^{\alpha\pm}) = e_p(f^\alpha/2) = e_p(n!/2) = e_p(|A_n|),$$

and the statement is proved for the case $[\tilde{\alpha}] = [\alpha] = [\alpha']$.

(ii) We assume now that $[\alpha] \neq [\tilde{\alpha}]$.

a) At first we would like to show that under the additional assumption $p \neq 2$ the restriction $[\alpha] \downarrow A_n$ cannot contain an irreducible constituent which belongs to a block of defect 0. If f^{α_A} denotes the dimension of such a supposed constituent of a block of defect 0, we have

$$e_p(f^{\alpha_A}) = e_p(n!/2) = e_p(n!) ,$$

the last equation is valid since we have assumed $p \neq 2$.

Because of $f^{\alpha_A} = f^{\alpha}$ or $= f^{\alpha}/2$ this implies

$$e_p(n!) = e_p(f^{\alpha_A}) = e_p(f^{\alpha}) \ ,$$

in contradiction to $[\alpha] \neq [\tilde{\alpha}]$.

b) If $p = 2$ we have for the dimension of a constituent $[\alpha]_A$ of $[\alpha] \downarrow A_n$ which belongs to a block of defect 0 and which is therefore modular irreducible:

$$e_2(f^{\alpha_A}) = e_2(n!/2) = e_2(n!) - 1 \ . \qquad (1)$$

With $e_2(f^{\alpha}) \geq e_2(f^{\alpha_A})$ we obtain

$$e_2(f^{\alpha}) = e_2(n!) - 1 \qquad (2)$$

since $[\alpha] \neq [\tilde{\alpha}]$.

Comparing (1) and (2) we get

$$e_2(f^{\alpha_A}) = e_2(f^{\alpha}) \ ,$$

such that because of $f^{\alpha_A} = f^{\alpha}$ or $= f^{\alpha}/2$ we obtain

$$[\alpha] \downarrow A_n = [\alpha]_A \ .$$

Since on account of (2) $[\alpha]$ contains exactly one 2-hook, $[\alpha]$ belongs to a 2-block corresponding to a 2-core with n-2 nodes. 2-cores are selfassociated: $[\tilde{\beta}] = [\tilde{\beta}']$ for every 2-core since for every 2-core $[\tilde{\beta}]$ we have

$$[\tilde{\beta}] = [0] \quad \text{or} \quad [\tilde{\beta}] = [\tilde{\beta}_1, \tilde{\beta}_1 - 1, \tilde{\beta}_1 - 2, \ldots, 2, 1] \ .$$

Hence the diagram $[\alpha]$ containing exactly one 2-hook is of one of the two following forms:

$$[\alpha] = [\tilde{\alpha}_1 + 2, \tilde{\alpha}_1 - 1, \ldots, 2, 1] \text{ or } [\alpha] = [\tilde{\alpha}_1, \tilde{\alpha}_1 - 1, \ldots, 2, 1^3].$$

These two representations are associated with respect to A_n, and their restrictions to A_n are equal. Hence only one of them yields a row of the decomposition matrix of A_n. Thus $[\alpha]_A$ forms its own block and this block consists actually of the constituents of restrictions $[\beta] \downarrow A_n$ with $[\tilde{\beta}] = [\tilde{\alpha}]$ or $[\tilde{\beta}] = [\tilde{\alpha}']$ since the last consideration shows that there is up to equivalence only one such constituent:

$$[\alpha]_A = [\tilde{\alpha}_1+2,\tilde{\alpha}_1-1,\ldots,2,1] \downarrow A_n = [\tilde{\alpha}_1,\tilde{\alpha}_1-1,\ldots,2,1^3] \downarrow A_n \ .$$

Thus the statement is proved in case that $[\alpha] \neq [\tilde{\alpha}]$ and $[\alpha]\downarrow A_n$ contains a constituent of a block of defect 0 and if $[\alpha] \neq [\tilde{\alpha}]$ this is only the case if $p = 2$.

c) Now let $[\beta]_A$ be an irreducible constituent of $[\beta] \downarrow A_n$ so that $[\tilde{\beta}] = [\tilde{\alpha}]$ or $= [\tilde{\alpha}']$.

a) and b) imply that we can assume that $[\beta]_A$ and the constituents $[\alpha]_A$ of $[\alpha] \downarrow A_n$ belong to blocks with defects >0. We have to show that $[\beta]_A$ and $[\alpha]_A$ belong to the same block. To prove this we use the class multipliers.

Since $[\beta]$ belongs to the S_n-block of $[\alpha]$ or of $[\alpha']$ we have (Curtis/Reiner [1], (85.12)):

$$\omega_a^\alpha \equiv \omega_a^\beta \quad (p)$$

for all the p-regular S_n-classes C^a which satisfy $C^a \subseteq A_n$.

We would like to show the validity of the analogous congruences for ω^{α_A} and ω^{β_A}, this would complete the proof of this part of

our statement (see Curtis/Reiner [1],(86.19)).
The theorem of Frobenius (4.55) implies that ζ^{α_A} resp. ζ^{β_A} agree with ζ^{α} resp. ζ^{β} or with $\frac{1}{2}\zeta^{\alpha}$ resp. $\frac{1}{2}\zeta^{\beta}$ on non-splitting classes of S_n. The same is valid for the dimensions, hence for such non-splitting classes $C^a \subseteq A_n$ we have also

$$\omega_a^{\alpha_A} \equiv \omega_a^{\beta_A} \quad (p) \ .$$

The elements of splitting classes consist of cycles of pairwise different lengths (see 1.23). Hence the order of the centralizer of a permutation which belongs to a splitting class is the product of its cycle lengths (see 2.32). If such a permutation is a p-regular one, p does not divide any one of these cycle lengths and therefore p doesn't divide the order of the centralizer such that the defect of such a p-regular splitting class is 0. Hence from a well known theorem (Curtis/Reiner [1], (86.27)) we get:

$$\omega_a^{\alpha_A} \equiv 0 \equiv \omega_a^{\beta_A} \quad (p)$$

for all the p-regular splitting classes, since $[\alpha]_A$ and $[\beta]_A$ belong to blocks of defects >0.

Thus for all the p-regular A_n-classes C^a we have

$$\omega_a^{\alpha_A} \equiv \omega_a^{\beta_A} \quad (p) \ ,$$

hence $[\alpha]_A$ and $[\beta]_A$ belong to the same block.

d) It remains to show, that the block of $[\alpha]_A$ does not contain a constituent $[\beta]_A$ of a representation $[\beta] \downarrow A_n$ with $[\widetilde{\beta}] \neq [\widetilde{\alpha}]$

and $[\tilde{\beta}] \neq [\tilde{\alpha}']$.

The ordinary irreducible representations belonging to a certain block B can be characterized as follows: two ordinary irreducible representations belong to B if and only if there exists a chain of ordinary irreducible representations beginning with one of the two representations and ending with the other one and so that modular representations associated to any two neighbours of the chain have an irreducible constituent in common.

Hence it suffices to show that modular representations $\overline{[\alpha]_A}$ and $\overline{[\beta]_A}$ associated with irreducible constituents $[\alpha]_A$ and $[\beta]_A$ of $[\alpha] \downarrow A_n$ and $[\beta] \downarrow A_n$ such that $[\tilde{\beta}] \neq [\tilde{\alpha}]$ and $[\tilde{\beta}] \neq [\tilde{\alpha}']$ have no irreducible constituent in common.

From Clifford's theory of representations of groups with normal divisors we get that an irreducible modular constituent F_A^i of $\overline{[\alpha]} \downarrow A_n$ is a constituent of the restriction $F_S^j \downarrow A_n$ of a modular irreducible representation F_S^j belonging either to the p-block of S_n with p-core $[\tilde{\alpha}]$ or to the associated p-block with p-core $[\tilde{\alpha}']$. But $[\beta]$ is <u>not</u> contained in one of these two blocks of S_n.

If now

$$\overline{[\beta]} \leftrightarrow \sum_k d_{ik} F_S^k$$

describes the modular decomposition of $[\beta]$, then $[\beta] \downarrow A_n$ has the same decomposition as the restriction of the modular representation on the right hand side:

$$\overline{[\beta]} \downarrow A_n \approx \sum_k d_{ik}(F_S^k \downarrow A_n) .$$

Thus the modular decomposition of this right hand side contains the decomposition of $\overline{[\beta]_A}$. But since $[\alpha]_A$ is contained in quite another block, this right hand side and hence the decomposition of $\overline{[\beta]_A}$, too, cannot contain a modular representation in common with $\overline{[\alpha]_A}$.

q.e.d.

7.2 and 7.6 enable us to evaluate the distribution of the ordinary irreducible representations of S_n and A_n into p-blocks in a very simple way and for every n and p.

But the problem of finding the decomposition numbers is still far from a satisfactory solution. We would like to describe this now. We shall treat the case p = 2 and report upon the known results. Thereafter we shall discuss what can be said about the decomposition matrix of A_n if the decomposition matrix of S_n is known. In certain cases, for example if $n \leq 2p$, it is easy to get the decomposition numbers of S_n (see Robinson [5], p. 122). But for the general case, there is only the method of explicitly reducing the representing matrices which is possible without a vast amount of calculations only for very small dimensions (see Robinson [5]). The use of an induction process seems to be much more promising.

As we soon shall see such a process together with some results on the decomposition numbers of special representations allows the evaluation of the decomposition numbers of S_n with respect to p=2 up to n = 9. We would like to show how this can be done.

For a fixed prime number p let

$$D^1 := D^1_{n,p} = (d^1_{ik})$$

denote the decomposition matrix of S_n with respect to p. (The upper index 1 will be shown to be of use in the following section where the generalized decomposition matrix will be considered whose first columns are built up by D^1.)

If the ordinary irreducible representation $[\alpha]_i$ of S_n belongs to the i-th row of D^1, i.e. if

$$\overline{[\alpha]_i} \leftrightarrow \sum_k d^1_{ik} F^k_S ,$$

the modular decomposition of $\sum_i d^1_{ik} [\alpha]_i$ is the same as the decomposition of the k-th principal indecomposable U_k of S_n:

7.7 $$U_k \approx \sum_i d^1_{ik}\overline{[\alpha]_i} .$$

The induced representation $U_k \uparrow S_{n+1}$ is a direct sum of principal indecomposables of S_{n+1}, hence $\sum_i d^1_{ik}(\overline{[\alpha]_i} \uparrow S_{n+1})$ has the same decomposition as a certain sum of principal indecomposables $\widetilde{U}_j$ of S_{n+1}:

7.8 $$\sum_i d^1_{ik}(\overline{[\alpha]_i} \uparrow S_{n+1}) \approx \sum_j a_j\widetilde{U}_j .$$

The branching theorem 4.52 implies, that the irreducible constituents of $[\alpha]_i \uparrow S_{n+1}$ are exactly the $[\beta]$ arising from $[\alpha]_i$ by adding one node. Restricting these additions in such a way that the arising diagrams $[\beta]$ have the same p-core, we can assure, that the decomposition of the arising representation of S_{n+1} contains principal indecomposables belonging to one block only.

7.9 Def.: The node in the i-th row and j-th column of $[\alpha]$ is called an r-node (with respect to p) if

$$j - i \equiv r \quad (p) .$$

By

$$[\alpha] \overset{r}{\uparrow} S_{n+1}$$

we shall denote the representation r-induced by $[\alpha]$ and consisting exactly of the representations $[\beta]$ (each with multiplicity 1) of S_{n+1} whose diagrams arise from $[\alpha]$ by adding an r-node.

Analogously we define the r-restriction of $[\alpha]$ to S_{n-1} and call the procedure the r-inducing resp. r-restricting process.

Using Nakayama's conjecture we obtain that the constituents of $[\alpha] \overset{r}{\uparrow} S_{n+1}$ have the same p-core so that they all belong to the same p-block of S_{n+1} and that all the diagrams with this p-core and out of $[\alpha] \uparrow S_{n+1}$ are contained in $[\alpha] \overset{r}{\uparrow} S_{n+1}$. Hence we have (see Robinson [5],6.11):

7.10 If $[\beta]$ and $[\gamma]$ arise from $[\alpha]$ by adding a node, then $[\widetilde{\beta}]=[\widetilde{\gamma}]$

<u>if and only if the added nodes are of the same residue class r modulo p.</u>

This implies for the decomposition matrix:

7.11 $$\sum_i d^1_{ik}(\overline{[\alpha]_i} \overset{r}{\uparrow} S_{n+1}) \approx \sum_j b_j \tilde{U}_j \ ,$$

<u>such that the $\tilde{U}_j$ with $b_j \neq 0$ belong to the same block of S_{n+1}.</u>

With this fundamental result we can start with our example p = 2. The only principal indecomposable of S_1 is $\overline{[1]}$, this is trivial. The residue class modulo 2 of the only node of the diagram [1] is 0. Thus

$$U_1 \overset{0}{\uparrow} S_2 \approx [1] \overset{0}{\uparrow} S_2 = \emptyset \ , \quad [1] \overset{1}{\uparrow} S_2 = [2]+[1^2] \approx U_1 \overset{1}{\uparrow} S_2 .$$

S_2 is a 2-group, hence it possesses only one principal indecomposable and the last equation together with 7.11 implies that this principal indecomposable has the same decomposition as $\overline{[2]} + \overline{[1^2]}$. Thus

7.12 $$D^1_{2,2} = \begin{bmatrix} 1 \\ 1 \end{bmatrix} \begin{matrix} [2] \\ [1^2] \end{matrix}$$

is the decomposition matrix of S_2 for p = 2.

To proceed with the r-inducing process we replace the nodes of the diagram by their residue classes modulo 2:

$$[2]: \ 0 \ 1 \qquad\qquad [1^2]: \ \begin{matrix} 0 \\ 1 \end{matrix} .$$

We obtain

$$U_1 \overset{0}{\uparrow} S_3 \approx [2] + [1^2] \overset{0}{\uparrow} S_3 = [3] + [1^3] ,$$
$$U_1 \overset{1}{\uparrow} S_3 \approx [2] + [1^2] \overset{1}{\uparrow} S_3 = 2[2,1] .$$

Nakayama's conjecture implies that [2,1] forms its own 2-block and is modular irreducible.

Besides this block S_3 possesses only one further principal indecomposable, hence this one has the same decomposition as $[3]+[1^3]$ and we obtain the decomposition matrix of S_3 for p = 2:

<u>7.13</u>
$$D^1_{3,2} = \begin{bmatrix} 1 & \\ 1 & \\ & 1 \end{bmatrix} \begin{matrix} [3] \\ [1^3] \\ [2,1] . \end{matrix}$$

(For the sake of simplicity we often omit the 0's.)

Then

$$[3]:\ 0\ 1\ 0\ , \quad [2,1]:\ \begin{matrix} 0\ 1 \\ 1\ \ \end{matrix}\ , \quad [1^3]:\ \begin{matrix} 0 \\ 1 \\ 0 \end{matrix}\ .$$

Thus

7.14
$$U_1 \overset{0}{\uparrow} S_4 \approx [3] + [1^3] \overset{0}{\uparrow} S_4 = \emptyset ,$$
$$U_1 \overset{1}{\uparrow} S_4 \approx [3] + [1^3] \overset{1}{\uparrow} S_4 = [4] + [3,1] + [2,1^2] + [1^4] ,$$
$$U_2 \overset{0}{\uparrow} S_4 \approx [2,1] \overset{0}{\uparrow} S_4 = [3,1] + [2^2] + [2,1^2] ,$$
$$U_2 \overset{1}{\uparrow} S_4 \approx [2,1] \overset{1}{\uparrow} S_4 = \emptyset .$$

S_4 has exactly two 2-regular classes so that $D^1_{4,2}$ consists of 2 columns.

7.14 implies, that the columns of the matrix

7.15
$$R_{4,2} := \begin{bmatrix} 1 & \\ 1 & 1 \\ 0 & 1 \\ 1 & 1 \\ 1 & \end{bmatrix} \begin{matrix} [4] \\ [3,1] \\ [2^2] \\ [2,1^2] \\ [1^4] \end{matrix}$$

are linear combinations (with nonnegatve integral coefficients) of the columns of $D^1_{4,2}$.

$R_{4,2}$ is a matrix of the same shape as $D^1_{4,2}$ and it is a lower triangular matrix. Hence its first column contains the first column of $D^1_{4,2}$ and its second column contains the second column of $D^1_{4,2}$. The triangular form implies furthermore, that the second column of $R_{4,2}$ cannot contain the first column of $D^1_{4,2}$ so that the second column of $R_{4,2}$ is equal to the second column of $D^1_{4,2}$, since decomposition numbers are integers. This second column of $R_{4,2}$ cannot be subtracted from the first column of $R_{4,2}$ without getting negative entries such that the first column of $R_{4,2}$ is equal to the first column of $D^1_{4,2}$:

<u>7.16</u>
$$D^1_{4,2} = \begin{bmatrix} 1 & \\ 1 & 1 \\ 0 & 1 \\ 1 & 1 \\ 1 & \end{bmatrix} \begin{matrix} [4] \\ [3,1] \\ [2^2] \\ [2,1^2] \\ [1^4] \end{matrix} .$$

Then

$$[4]: 0\;1\;0\;1 \;,\quad [3,1]: \begin{matrix} 0 & 1 & 0 \\ 1 & & \end{matrix} \;,\quad [2^2]: \begin{matrix} 0 & 1 \\ 1 & 0 \end{matrix} \;,\quad [2,1^2]: \begin{matrix} 0 & 1 \\ 1 & \\ 0 & \end{matrix} \;,\quad [1^4]: \begin{matrix} 0 \\ 1 \\ 0 \\ 1 \end{matrix}$$

And we obtain:

7.17

$$[4]+[3,1]+[2,1^2]+[1^4] \overset{0}{\uparrow} S_5 = [5]+[3,2]+2[3,1^2]+[2^2,1]+[1^5] \quad (1)$$

$$'' \overset{1}{\uparrow} S_5 = 2([4,1]+[2,1^3]) \quad (2)$$

$$[3,1]+[2^2]+[2,1^2] \overset{0}{\uparrow} S_5 = 3([3,2]+[3,1^2]+[2^2,1]) \quad (3)$$

$$'' \overset{1}{\uparrow} S_5 = [4,1]+[2,1^3] \quad (4)$$

Nakayama's conjecture implies, that [4,1] forms a 2-block together with $[2,1^3]$ so that we obtain from (4), that $\overline{[4,1]}+\overline{[2,1^3]}$ has the same decomposition as a principal indecomposable of S_5 since this block contains only one principal indecomposable (the number of modular irreducible representations in a given block of S_n is known).

There are two further principal indecomposables, hence 7.17 (1) and (3) suggest that we investigate, how the matrix

7.18

$$R_{5,2} := \begin{bmatrix} 1 & & \\ 1 & 2 & \\ 2 & 2 & \\ 1 & 2 & \\ 1 & & \\ & & 1 \\ & & 1 \end{bmatrix} \begin{matrix} [5] \\ [3,2] \\ [3,1^2] \\ [2^2,1] \\ [1^5] \\ [4,1] \\ [2,1^3] \end{matrix}$$

should be corrected to yield $D^1_{5,2}$.

Here as in the following examples we observe, that this matrix $R_{n,p}$ which we would like to correct to obtain $D^1_{n,p}$, has the following important properties:

(i) $R_{n,p}$ is a direct sum of submatrices of the same shape as the submatrices of $D^1_{n,p}$ (there is a general formula for the number of modular irreducible representations in a given block of S_n (see Robinson [5],6.37) so that $R_{n,p}$ has the same number of columns as $D^1_{n,p}$.

(ii) The columns of $R_{n,p}$ are rational integral combinations (with nonnegative coefficients)of the columns of $D^1_{n,p}$.

(iii)The submatrices of which $R_{n,p}$ is a direct sum are lower triangular matrices with nonzero entries along the leading diagonal.

Let us now assume that we have obtained by r-inducing on $D^1_{n-1,p}$ a matrix $R_{n,p}$ which satisfies (i)-(iii). We have to describe what can be done to get $D^1_{n,p}$ from $R_{n,p}$.

We fix attention on one of the summands of $R_{n,p}$ (see (i)). Since (iii) is valid this summand is a lower triangular matrix.This implies that for a certain rearrangement of the columns of $D^1_{n,p}$ the corresponding submatrix of $D^1_{n,p}$ is also lower triangular. We assume, that this rearrangement has been carried out so that it remains to obtain from the i-th column of this summand of $R_{n,p}$ the i-th column of the corresponding summand of $D^1_{n,p}$ (we can be sure, that the i-th column of the summand of $R_{n,p}$ includes the i-th column of the summand of $D^1_{n,p}$ at least once and this wanted i-th column of $D^1_{n,p}$ is not included in a column farther to the right in $R_{n,p}$).

Hence the column to the extreme right of the considered summand of $R_{n,p}$ (cf. example 7.18) is an integral multiple of the corresponding column of $D^1_{n,p}$. Thus we have to check whether this column can be divided by a natural number to yield a new column over $\mathbb{Z}$.

7.19 <u>If the column to the extreme right of the considered summand of $R_{n,p}$ can be divided by a natural number, this division has to be carried out to yield the corresponding column of $D^1_{n,p}$.</u>

Proof: We know that the Cartan matrix

$$C^1_{n,p} := {}^tD^1_{n,p}D^1_{n,p}$$

(${}^tD^1_{n,p}$ the transposed matrix) has a determinant whose value is a power of p.

Furthermore if $D^{1*}_{n,p}$ arises from $D^1_{n,p}$ by adding a column of $D^1_{n,p}$ to another column of $D^1_{n,p}$, it is easy to see that

$$\det({}^tD^{1*}_{n,p}D^{1*}_{n,p}) = \det C^1_{n,p} .$$

With this in mind it is not too difficult to check, that

$$\det({}^tR_{n,p}R_{n,p}) = r_1^2 \ldots r_k^2 \det C^1_{n,p} , \tag{1}$$

if $r_i \in \mathbb{N}$ is the multiplicity of the i-th column of $D^1_{n,p}$ in the i-th column of $R_{n,p}$ (after the suitable rearrangement of the columns of $D^1_{n,p}$ has been carried out).

Finally we recall that $D^1_{n,p}$ has maximal p-rank (see Curtis/Reiner [1],(83.5)).

Hence if it is possible we have to divide the considered column to the extreme right by p. That we have to divide these entries by other common factors we see from (1), recalling that $\det C^1_{n,p}$ is a power of p.

q.e.d.

Thus we obtain one column of $D^1_{n,p}$ anyway.
For example the second column of 7.18 has to be divided by 2: In

7.20
$$R^*_{5,2} := \begin{bmatrix} 1 & & \\ 1 & 1 & \\ 2 & 1 & \\ 1 & 1 & \\ 1 & & \\ & & 1 \\ & & 1 \end{bmatrix} \begin{matrix} [5] \\ [3,2] \\ [3,1^2] \\ [2^2,1] \\ [1^5] \\ [4,1] \\ [2,1^3] \end{matrix}$$

at least the second and third columns agree with the corresponding columns of $D^1_{5,2}$.
It remains to decide whether the second column has to be subtracted from the first one in $R^*_{5,2}$ to obtain $D^1_{5,2}$.
This can be decided using a result of M.H. Peel on the decomposition numbers of $\overline{[n-2,2]}$.
Before we mention Peel's result we give the other general result on decomposition numbers of S_n (Farahat [1], cf. also Kerber [1]) and which is of equal importance:

<u>7.21</u> <u>If p does not divide n, then $\overline{[n-1,1]}$ is irreducible.</u>
<u>If p divides n and n>2, then there are exactly 2 irreducible constituents of $\overline{[n-1,1]}$, each of them with multiplicity 1, and one of them is the identity representation.</u>

Peel's result (Peel [1]) reads as follows:

7.22 1) If p doesnot divide n-2 we distinguish the cases p=2 and $p \neq 2$:

a) In case that $p \neq 2$ and

(i) $p \nmid n-1$, then $\overline{[n-2,2]}$ is irreducible. If

(ii) $p \mid n-1$, then $\overline{[n-2,2]}$ has exactly two different irreducible constituents, each with multiplicity 1, one of them is the identity representation.

b) If p=2 and

(i) n=2a+1, $2 \nmid a$, $\overline{[n-2,2]}$ is irreducible. If

(ii) n=2a+1, $2 \mid a$, then $\overline{[n-2,2]}$ contains exactly two irreducible constituents, each with multiplicity 1, one of them is the identity representation.

2) If on the other hand p divides n-2, and

a) p=2 we have to distinguish the following two cases:

(i) If n=2a, $2 \mid a$ and a>2 ($\overline{[2]}$ and $\overline{[2^2]}$ are irreducible as we have seen above), then $\overline{[n-2,2]}$ possesses exactly two different irreducible constituents, each one with multiplicity 1, one of them is of dimension n-2.

(ii) If n=2a, $2 \nmid a$, then $\overline{[n-2,2]}$ contains exactly three different irreducible constituents, each one with multiplicity 1, one of them is the identity representation, another one is of dimension n-2.

b) If $p \neq 2$, $\overline{[n-2,2]}$ possesses exactly 2 different irreducible constituents, every one with multiplicity 1, one of them is the irreducible representation $\overline{[n-1,1]}$ (cf. 7.21).

From 7.22 1) b) (ii) we obtain with 7.20:

7.23
$$D^1_{5,2} = \left[\begin{array}{cc|c} 1 & & \\ 1 & 1 & \\ 2 & 1 & \\ 1 & 1 & \\ 1 & & \\ \hline & & 1 \\ & & 1 \end{array}\right] \begin{array}{l} [5] \\ [3,2] \\ [3,1^2] \\ [2^2,1] \\ [1^5] \\ [4,1] \\ [2,1^3] \end{array} .$$

The r-inducing process, Nakayama's conjecture and 7.21/7.22 yield

7.24
$$D^1_{6,2} = \left[\begin{array}{ccc|c} 1 & & & \\ 1 & 1 & & \\ 1 & 1 & 1 & \\ 2 & 1 & 1 & \\ 1 & 0 & 1 & \\ 1 & 0 & 1 & \\ 2 & 1 & 1 & \\ 1 & 1 & 1 & \\ 1 & 1 & & \\ 1 & & & \\ \hline & & & 1 \end{array}\right] \begin{array}{l} [6] \\ [5,1] \\ [4,2] \\ [4,1^2] \\ [3^2] \\ [2^3] \\ [3,1^3] \\ [2^2,1^2] \\ [2,1^4] \\ [1^6] \\ [3,2,1] \end{array} .$$

The application of the r-inducing process to $D^1_{6,2}$ yields the following matrix (we used 7.19 and 7.22):

7.25 $$R_{7,2} := \left[\begin{array}{ccc|cc} 1 & & & & \\ 0 & 1 & & & \\ 1 & 1 & & & \\ 1 & 1 & 1 & & \\ 1 & 0 & 1 & & \\ 1 & 0 & 1 & & \\ 1 & 1 & 1 & & \\ 1 & 1 & & & \\ 0 & 1 & & & \\ 1 & & & & \\ \hline & & & 1 & \\ & & & 1 & 1 \\ & & & 2 & 1 \\ & & & 1 & 1 \\ & & & 1 & \end{array}\right] \begin{array}{l} [7] \\ [5,2] \\ [5,1^2] \\ [4,2,1] \\ [3^2,1] \\ [3,2^2] \\ [,2,1^2] \\ [3,1^4] \\ [2^2,1^3] \\ [1^7] \\ [6,1] \\ [4,3] \\ [4,1^3] \\ [2^3,1] \\ [2,1^5] \end{array}$$

To check, that this matrix is the decomposition matrix of S_7 we use a nice trick: 7.24 implies that the difference of the characters of $\overline{[3^2]}$ and $\overline{[6]}$ is the Brauer character of F_3, the third 2-modular irreducible representation of S_6 which belongs to the third column of $D^1_{6,2}$. For short:

$$F_3 = \overline{[3^2]} - \overline{[6]} .$$

Hence we may induce to get

7.26 $$F_3 \uparrow S_7 \approx (\overline{[3^2]}-\overline{[6]}) \uparrow S_7 \approx \overline{[3^2]} \uparrow S_7 - \overline{[6]} \uparrow S_7$$
$$\approx \overline{[4,3]} + \overline{[3^2,1]} - \overline{[7]} - \overline{[6,1]} .$$

Now we look at this decomposition and 7.25.

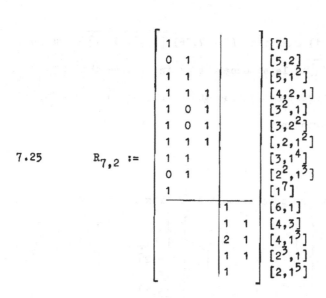

The right hand side of 7.26 contains each 2-modular irreducible representation of S_7 with nonnegative multiplicity. From 7.25 we know, that $\overline{[7]}$ as well as $\overline{[6,1]}$ are irreducible representations (see also 7.21) such that (see 7.26) $\overline{[4,3]}+\overline{[3^2,1]}$ contains $\overline{[7]}$ as well as $\overline{[6,1]}$ at least once. Hence in 7.25 neither the third column can be subtracted from the first one nor the fifth column from the fourth one:

<u>7.27</u> $$R_{7,2} = D^1_{7,2} .$$

Applying the r-inducing process to this matrix we obtain:

7.28
$$R_{8,2} := \left[\begin{array}{ccccc|c}
1 & & & & & \\
1 & 1 & & & & \\
0 & 1 & 1 & & & \\
1 & 1 & 1 & & & \\
0 & 1 & 1 & 1 & & \\
1 & 2 & 1 & 1 & & \\
0 & 1 & 0 & 1 & & \\
2 & 1 & 1 & 1 & 1 & \\
2 & 0 & 1 & 0 & 1 & \\
2 & 2 & 2 & 1 & 1 & \\
1 & 2 & 1 & 1 & 0 & \\
2 & 0 & 0 & 0 & 1 & \\
2 & 0 & 1 & 0 & 1 & \\
2 & 1 & 1 & 1 & 1 & \\
0 & 1 & 0 & 1 & & \\
0 & 1 & 1 & 1 & & \\
1 & 1 & 1 & & & \\
0 & 1 & 1 & & & \\
1 & 1 & & & & \\
1 & & & & & \\
\hline
 & & & & & 1 \\
 & & & & & 1
\end{array}\right]
\begin{array}{l}
[8] \\ [7,1] \\ [6,2] \\ [6,1^2] \\ [5,3] \\ [5,1^3] \\ [4^2] \\ [4,3,1] \\ [4,2^2] \\ [4,2,1^2] \\ [4,1^4] \\ [3^2,2] \\ [3^2,1^2] \\ [3,2^2,1] \\ [2^4] \\ [2^3,1^2] \\ [3,1^5] \\ [2^2,1^4] \\ [2,1^6] \\ [1^8] \\ [5,2,1] \\ [3,2,1^3]
\end{array} .$$

7.21 and 7.22 imply, that at most the fifth column of this matrix has to be subtracted from the first one and/or the fourth column from the second one.

Once more we consider an induced character:

7.29 $$F_5 \uparrow S_8 \approx (\overline{[4,3]}-\overline{[6,1]}) \uparrow S_8$$
$$\approx \overline{[5,3]}+\overline{[4^2]}+\overline{[4,3,1]}-\overline{[7,1]}-\overline{[6,2]}-\overline{[6,1^2]} .$$

Since the first three rows of $R_{8,2}$ agree with the first three rows of $D^1_{8,2}$ (cf. 7.21/7.22), $\overline{[4,3,1]}$ contains at least twice the irreducible representation $\overline{[8]}$ as well as $\overline{[5,3]}+\overline{[4^2]}+\overline{[4,3,1]}$ contains at least thrice the second irreducible representation $\overline{[7,1]}-\overline{[8]}$. This implies

7.30 $$R_{8,2} = D^1_{8,2} .$$

r-inducing on the columns of $D^1_{8,2}$ yields some possibilities for the columns of $D^1_{9,2}$ (see 7.32 on page 154) which we would like to investigate now.

We consider at first the lower right hand submatrix for to obtain the box of $D^1_{9,2}$ which includes $\overline{[8,1]}$.

That the column to the extreme right has to be divided by 3 we obtain from 7.19.

Since the first entry of the second possibility for the first column is 1, the first possibility for this column contains twice the first column of the corresponding box of $D^1_{9,2}$:

7.31
$$\begin{matrix}1\\0\\1\\3\\3\\3\\1\\3\\0\\1\end{matrix}$$

contains the sixth column of $D^1_{9,2}$.

7.32

1											[9]
1	2	1									[7,2]
2	2	1									$[7,1^2]$
1	2	2	1								[6,2,1]
0	2	1	0	2							[5,4]
2	2	2	1	2	1						[5,3,1]
2	0	1	1	0	1						$[5,2^2]$
3	4	3	1	2	1						$[5,2,1^2]$
2	4	2	0	2	0						$[5,1^4]$
2	2	1	0	2	1						$[4^2,1]$
3	4	3	1	2	1						$[4,2,1^3]$
2	0	0	0	0	1						$[3^3]$
2	0	1	1	0	1						$[3^2,1^3]$
2	2	1	0	2	1						$[3,2^3]$
2	2	2	1	2	1						$[3,2^2,1^2]$
1	2	2	1	0							$[3,2,1^4]$
2	2	1		0							$[3,1^6]$
0	2	1		2							$[2^4,1]$
1	2	1									$[2^2,1^5]$
1											$[1^9]$
						2	1				[8,1]
						0	2	2	1		[6,3]
						2	3	2	1		$[6,1^3]$
						6	1	2	1	3	[4,3,2]
						6	3	4	2	3	$[4,3,1^2]$
						6	3	4	2	3	$[4,2^2,1]$
						2	3	2	1	0	$[4,1^5]$
						6	1	2	1	3	$[3^2,2,1]$
						0	2	2	1		$[2^3,1^3]$
						2	1				$[2,1^7]$.

That 7.31contains the sixth column of $D^1_{9,2}$ implies, that the second one of the two possibilities for this sixth column given in 7.32 contains at least twice the seventh column of $D^1_{9,2}$. Since 7.31 this would be impossible unless the column

7.33
$$\begin{array}{c}1\\1\\1\\2\\2\\1\\1\\1\end{array}$$

would contain the eigth column of $D^1_{9,2}$. This implies, that

7.34
$$\begin{array}{cc}1&\\1&\\0&1\\1&1\\1&1\\1&0\\0&1\\1&\end{array}$$

are the two columns to the extreme right of $D^1_{9,2}$.

Hence there remains to examine the following two possibilities for the sixth column and the corrected seventh and eigth column:

7.35
$$\left[\begin{array}{cc|cc}1&1&&\\0&2&1&\\1&3&1&\\3&1&0&1\\3&3&1&1\\3&3&1&1\\1&3&1&0\\3&1&0&1\\0&2&1&\\1&1&&\end{array}\right]$$

The combination of these two possibilities for the first column gives, that

7.36
$$\begin{array}{c}1\\0\\1\\1\\1\\1\\1\\1\\0\\1\end{array}$$

contains the first column of the corresponding box of $D^1_{9,2}$.

Thus it remains to decide whether

$$7.37 \qquad \left[\begin{array}{ccc} 1 & & \\ 0 & 1 & \\ 1 & 1 & \\ 0 & 0 & 1 \\ 0 & 1 & 1 \\ 0 & 1 & 1 \\ 1 & 1 & 0 \\ 0 & 0 & 1 \\ 0 & 1 & \\ 1 & & \end{array}\right] \begin{array}{l} [8,1] \\ [6,3] \\ [6,1^3] \\ [4,3,2] \\ [4,3,1^2] \\ [4,2^2,1] \\ [4,1^5] \\ [3^2,2,1] \\ [2^3,1^3] \\ [2,1^7] \end{array}$$

can be a submatrix of $D^1_{9,2}$ or not.

To decide this we consider again a special induced character:

$$7.38 \qquad F_5 \uparrow S_9 \approx (\overline{[3^2,2]}-2\overline{[8]}) \uparrow S_9$$
$$\approx \overline{[4,3,2]}+\overline{[3^3]}+\overline{[3^2,2,1]}-2\overline{[9]}-2\overline{[8,1]} .$$

This decomposition implies, that $\overline{[4,3,2]}+\overline{[3^2,2,1]}$ contains at least twice the irreducible constituent $\overline{[8,1]}$ so that 7.36 describes the first column of the considered submatrix of $D^1_{9,2}$.

It remains to investigate the submatrix of 7.32 which contains the identity representation.

That the only possibility for the fourth column has to be divided by 2 we obtain from 7.19 since the fifth column cannot be subtracted. Hence 7.31 suggests to examine the matrix 7.39 (see the following page), whose i-th column contains the i-th column of $D^1_{9,2}$ $(1 \leq i \leq 5)$.

Obviously it remains to check the first and the second column of 7.39. The other columns are columns of $D^1_{9,2}$ as they stand since from these columns not any other one can be subtracted without yielding negative entries.

7.39
$$\left[\begin{array}{ccccc}
1 & & & & \\
1 & 1 & & & \\
2 & 1 & & & \\
1 & 1 & 1 & & \\
0 & 1 & 0 & 1 & \\
2 & 1 & 1 & 1 & 1 \\
2 & 0 & 1 & 0 & 1 \\
3 & 2 & 1 & 1 & 1 \\
2 & 2 & 0 & 1 & 0 \\
2 & 1 & 0 & 1 & 1 \\
3 & 2 & 1 & 1 & 1 \\
2 & 0 & 0 & 0 & 1 \\
2 & 0 & 1 & 0 & 1 \\
2 & 1 & 0 & 1 & 1 \\
2 & 1 & 1 & 1 & 1 \\
1 & 1 & 1 & 0 & \\
2 & 1 & & 0 & \\
0 & 1 & & 1 & \\
1 & 1 & & & \\
1 & & & &
\end{array}\right]
\begin{array}{l}
[9] \\ [7,2] \\ [7,1^2] \\ [6,2,1] \\ [5,4] \\ [5,3,1] \\ [5,2^2] \\ [5,2,1^2] \\ [5,1^4] \\ [4^2,1] \\ [4,2,1^3] \\ [3^3] \\ [3^2,1^3] \\ [3,2^3] \\ [3,2^2,1^2] \\ [3,2,1^4] \\ [3,1^6] \\ [2^4,1] \\ [2^2,1^5] \\ [1^9]
\end{array}$$

To examine the first and second column of this matrix we consider

7.40
$$F_4 \uparrow S_9 \approx (\overline{[8]}+\overline{[5,3]}-\overline{[6,1^2]}) \uparrow S_9$$
$$\approx \overline{[9]}+\overline{[8,1]}+\overline{[6,3]}+\overline{[5,4]}+\overline{[5,3,1]}-\overline{[7,1^2]}-\overline{[6,2,1]}-\overline{[6,1^3]} .$$

Since the first three rows of 7.39 are correct as they stand (cf. 7.22), the representation $\overline{[7,1^2]}+\overline{[6,2,1]}$ contains the second irreducible representation $\overline{[7,2]}-\overline{[9]}$ with multiplicity 2 (notice that the third column of 7.39 cannot be subtracted from the second one). Hence 7.40 implies that $\overline{[5,4]}+\overline{[5,3,1]}$ contains $\overline{[7,2]}-\overline{[9]}$ at least twice such that neither the fourth nor the fifth column has to be subtracted from the second column of 7.39.

The second column of 7.39 is therefore correct as it stands.

It remains to consider the first column and to decide whether the third and/or fifth column have to be subtracted or not.

The first three entries of this column are correct as they stand.

We consider the decomposition

7.41 $F_3 \uparrow S_9 \approx (\overline{[8]}+\overline{[6,2]}-\overline{[7,1]}) \uparrow S_9$

$\approx \overline{[9]}+\overline{[8,1]}+\overline{[7,2]}+\overline{[6,3]}+\overline{[6,2,1]}-\overline{[8,1]}-\overline{[7,2]}-\overline{[7,1^2]}$.

$\overline{[8,1]}+\overline{[7,2]}+\overline{[7,1^2]}$ contains thrice the irreducible representation $\overline{[9]}$, hence at least this is valid for $\overline{[9]}+\overline{[7,2]}+\overline{[6,2,1]}$. Thus the third column has not to be subtracted from the first one so that even the first four entries of this column agree with $D^1_{9,2}$.

With this we return to 7.40: $\overline{[7,1^2]}+\overline{[6,2,1]}$ contains thrice the irreducible constituent $\overline{[9]}$ such that this representation is contained in $\overline{[9]}+\overline{[5,3,1]}$ at least with multiplicity 3. Hence also the fifth column has not to be subtracted from the first one.

Hence 7.42 (on page 159) is the decomposition matrix of S_9 with respect to p=2.

Thus we have verified without any an explicit reduction of a representation the decomposition numbers of S_n for p=2 and $n \leq 9$ which Robinson gave (Robinson [5],[6]). We have shown, how a combination of the r-inducing process together with a use of Farahat's and Peel's results yields these far reaching results. But it

7.42

1								[9]
1	1							[7,2]
2	1							$[7,1^2]$
1	1	1						[6,2,1]
0	1	0	1					[5,4]
2	1	1	1	1				[5,3,1]
2	0	1	0	1				$[5,2^2]$
3	2	1	1	1				$[5,2,1^2]$
2	2	0	1	0				$[5,1^4]$
2	1	0	1	1				$[4^2,1]$
3	2	1	1	1				$[4,2,1^3]$
2	0	0	0	1				$[3^3]$
2	0	1	0	1				$[3^2,1^3]$
2	1	0	1	1				$[3,2^3]$
2	1	1	1	1				$[3,2^2,1^2]$
1	1	1	0					$[3,2,1^4]$
2	1		0					$[3,1^6]$
0	1		1					$[2^4,1]$
1	1							$[2^2,1^5]$
1								$[1^9]$
					1			[8,1]
					0	1		[6,3]
					1	1		$[6,1^3]$
					1	0	1	[4,3,2]
					1	1	1	$[4,3,1^2]$
					1	1	1	$[4,2^2,1]$
					1	1	0	$[4,1^5]$
					1	0	1	$[3^2,2,1]$
					0	1		$[2^3,1^3]$
					1			$[2,1^7]$

should be mentioned, that in default of further general results like 7.21/7.22 we cannot proceed further for p = 2. For the evaluation of the decomposition matrix of S_{10} we get by r-inducing on 7.42 the submatrix 7.43 of $R_{10,2}$ which contains the submatrix of $D^1_{10,2}$ which includes $\overline{[10]}$:

$$
7.43 \qquad \begin{bmatrix}
1 &&&&&& \\
1 & 1 &&&&& \\
1 & 1 & 1 &&&& \\
2 & 1 & 1 &&&& \\
1 & 0 & 1 & 1 &&& \\
2 & 1 & 1 & 1 &&& \\
0 & 0 & 1 & 1 & 1 && \\
3 & 0 & 2 & 1 & 1 & 1 & \\
3 & 0 & 1 & 0 & 0 & 1 & \\
4 & 1 & 3 & 1 & 1 & 1 & \\
2 & 1 & 2 & 1 & 1 & 0 & \\
0 & 0 & 1 & 0 & 1 & 0 & \\
4 & 1 & 1 & 0 & 1 & 1 & 1 \\
5 & 1 & 3 & 1 & 2 & 1 & 1 \\
5 & 1 & 2 & 1 & 1 & 1 & 1 \\
2 & 1 & 2 & 1 & 1 & 0 & 0 \\
2 & 1 & 1 & 0 & 1 & 0 & 1 \\
2 & 1 & 1 & 1 & 1 & 0 & 1 \\
2 & 1 & 0 & 0 & 0 & 0 & 1 \\
5 & 1 & 2 & 1 & 1 & 1 & 1 \\
2 & 1 & 1 & 1 & 1 & 0 & 1 \\
5 & 1 & 3 & 1 & 2 & 1 & 1 \\
4 & 1 & 3 & 1 & 1 & 1 & 0 \\
2 & 1 & 1 & 1 & 0 & 0 & 0 \\
2 & 1 & 0 & 0 & 0 & 0 & 1 \\
2 & 1 & 1 & 0 & 1 & 0 & 1 \\
4 & 1 & 1 & 0 & 1 & 1 & 1 \\
3 & 0 & 1 & 0 & 0 & 1 & \\
3 & 0 & 2 & 1 & 1 & 1 & \\
2 & 1 & 1 & 0 & 0 && \\
0 & 0 & 1 & 0 & 1 && \\
0 & 0 & 1 & 1 & 1 && \\
1 & 0 & 1 & 1 &&& \\
1 & 1 & 1 &&&& \\
1 & 1 &&&&& \\
1 &&&&&&
\end{bmatrix}
$$

(cf. the table 2-10 in Robinson [5], Appendix, which has to be corrected). The last for columns of 7.43 agree with the last four columns of the corresponding submatrix of $D^1_{10,2}$.

These methods have been used also for the case p=3, where the first difficulty arises at n=8. The question is whether $\overline{[5,3]}$ contains $\overline{[8]}$ or not. The answer is negative: $\overline{[5,3]}$ does not contain $\overline{[8]}$ and using this result, the decomposition matrices have been calculated up to n=10 (Kerber/Peel [1]). The reader can find there a necessary and sufficient condition that $\overline{[n-3,3]}$ contains

$\overline{[n]}$. For further interesting results especially on the decomposition of hook-representations $\overline{[n-r,1^r]}$ the reader is referred to Peel [2].

Concluding these considerations of the decomposition numbers of S_n and gathering up our experiences we dare to give a conjecture:

__7.44 Conjecture:__ The submatrices of which the decomposition matrix $D^1_{n,p}$ of S_n for p is a direct sum are for a suitable rearrangement of the columns lower triangular matrices with 1's along the leading diagonal, if the first rows of the considered submatrix correspond to diagrams with no p rows of equal length in their natural order.

Parts of this conjecture but not the full statement have been proved by Robinson and O.E. Taulbee (Robinson [5], Taulbee [1]).

Concluding this section we would like to investigate what can be said about the decomposition matrix $D^1_{A_n,p}$ of A_n if $D^1_{S_n,p}$ is known. 7.6 yields the distribution of the ordinary irreducible representations of A_n into p-blocks. Hence for to obtain $D^1_{A_n,p}$ it remains to describe how we can get its columns from the columns of $D^1_{S_n,p}$ and to describe what happens with the corresponding entry of $D^1_{S_n,p}$, if the representations are restricted to A_n.

As in the ordinary case we can apply Clifford's theory of representations of groups with normal divisors. Our aim is to describe modular irreducible representations F_S of S_n whose restriction is reducible resp. irreducible.

From Clifford's theory we conclude that the irreducible representations of S_n over an algebraically closed field K (of any characteristic) can be obtained in the following way: Take an irreducible representations F_A of A_n and find its inertia group. Since $|S_n:A_n|\leq 2$ this inertia group is A_n or S_n. If A_n is the inertia group, then $F_A \uparrow S_n$ is an irreducible representation of S_n. If S_n is the inertia group of F_A, the F_A can be extended to an irreducible representation $\tilde{F}_A$ of S_n, and with this representation we can construct two irreducible representations of S_n:

$$\tilde{F}_A = \tilde{F}_A \otimes \overline{[2]} \quad , \quad \tilde{F}_A \otimes \overline{[1^2]}$$

(which need not be different, e.g. if char K = 2: $\overline{[2]} = \overline{[1^2]}$, and hence $\tilde{F}_A = \tilde{F}_A \otimes \overline{[1^2]}$).

And in this way we obtain all the irreducible representations of S_n.

This implies:

7.45 <u>If F_S is an irreducible 2-modular representation of S_n (n>1), then the following is valid:</u>

(i) $F_S \downarrow A_n$ reducible $\Leftrightarrow \exists\, F_A$: $F_S = F_A \uparrow S_n$.

(ii) $F_S \not\sim F_S \otimes \overline{[1^n]} \Rightarrow F_S \downarrow A_n = (F_S \otimes \overline{[1^n]}) \downarrow A_n$ irreducible.

If $\alpha=\alpha'$, then $[\alpha]\downarrow A_n$ splits into two mutually conjugate and irreducible representations of A_n:

$$[\alpha]\downarrow A_n = [\alpha]^+ + [\alpha]^-$$

so that

$$[\alpha]^{+(a)} \sim [\alpha]^- \text{ , } \forall\, a \in S_n \backslash A_n.$$

Thus also

$$\overline{[\alpha]^{+}}^{(a)} \sim \overline{[\alpha]^-}$$

and hence the constituents of $\overline{[\alpha]^-}$ are conjugates of the constituents of $\overline{[\alpha]^+}$.

This implies

<u>7.46</u> <u>If $F_S \sim F_S \otimes \overline{[1^n]}$ and the multiplicity of F_S in $\overline{[\alpha]}$ is odd and $\alpha = \alpha'$, then $F_S\downarrow A_n$ is reducible.</u>

Suppose now that using 7.46 we have succeeded in picking out the columns of $D^1_{S_n,p}$ which belong to modular irreducible representations F_S whose restriction to A_n is reducible (i.e. F_S is self-associated with respect to A_n). Then under certain circumstances (which are fulfilled in all the known cases $D^1_{S_n,p}$) we are able to evaluate $D^1_{A_n,p}$ at once.

Let us consider the row of $D^1_{S_n,p}$ which belongs to $[\alpha]$ and the column which belongs to F_S. We denote by a resp. b the multiplicity of F_S resp. $F_S \otimes \overline{[1^n]}$ in $\overline{[\alpha]}$ such that $D^1_{S_n,p}$ contains the following submatrix:

$$\begin{array}{ccccc} & & F_S & & F_S \otimes \overline{[1^n]} \\ & & \vdots & & \vdots \\ [\alpha] & \cdots & a & \cdots & b \\ & & \vdots & & \vdots \\ [\alpha'] & \cdots & b & \cdots & a \end{array}$$

(i) If $\alpha \neq \alpha'$, $F_S \nsim F_S \otimes \overline{[1^n]}$:

In this case the row of $[\alpha']$ and the column of $F_S \otimes \overline{[1^n]}$ has to be cancelled, and in the row of $[\alpha] \downarrow A_n$ we have in the column of $F_S \downarrow A_n$ the decomposition number a+b, since

$$\overline{[\alpha]} \leftrightarrow aF_S + b(F_S \otimes \overline{[1^n]}) + \ldots$$
$$\Rightarrow \overline{[\alpha]} \downarrow A_n \leftrightarrow a(F_S \downarrow A_n) + b(F_S \otimes \overline{[1^n]} \downarrow A_n) + \ldots$$
$$= (a+b)F_S \downarrow A_n + \ldots .$$

(ii) $\alpha \neq \alpha'$, $F_S \sim F_S \otimes \overline{[1^n]}$, $F_S \downarrow A_n$ irreducible:

It is trivial, that a is the multiplicity of $F_S \downarrow A_n$ in $\overline{[\alpha]} \downarrow A_n$.

(iii) $\alpha \neq \alpha'$, $F_S \downarrow A_n$ reducible ($\Rightarrow F_S \sim F_S \otimes \overline{[1^n]}$):

Then $F_S \downarrow A_n \leftrightarrow F_S^+ + F_S^-$ with two mutually conjugate and irreducible representations $F_S^{\pm}$ of A_n. Obviously $\overline{[\alpha]} \downarrow A_n$ contains F_S^+ as well as F_S^- with multiplicity a.

(iv) $\alpha = \alpha'$, $F_S \nsim F_S \otimes \overline{[1^n]}$:

a = b is the multiplicity of $F_S \downarrow A_n$ in $\overline{[\alpha]^+}$ as well as in $\overline{[\alpha]^-}$

since $F_S \downarrow A_n$ is selfconjugate.

(v) $\alpha = \alpha'$, $F_S \sim F_S \otimes \overline{[1^n]}$, $F_S \downarrow A_n$ irreducible:

$F_S \downarrow A_n$ is selfconjugate again. Thus $a = b$ is even and $a/2$ is the multiplicity of $F_S \downarrow A_n$ in $\overline{[\alpha]^+}$ as well as in $\overline{[\alpha]^-}$.

(vi) $\alpha = \alpha'$, $F_S \downarrow A_n \leftrightarrow F_S^+ + F_S^-$:

The theory provides an answer only if a = 0 or a = 1. In this case we have the submatrices

$$\begin{array}{cc} & \begin{array}{cc} F_S^+ & F_S^- \end{array} \\ \begin{array}{c} [\alpha]^+ \\ [\alpha]^- \end{array} & \begin{bmatrix} 0 & 0 \\ 0 & 0 \end{bmatrix} \end{array} \quad \text{or} \quad \begin{array}{cc} & \begin{array}{cc} F_S^+ & F_S^- \end{array} \\ \begin{array}{c} [\alpha]^+ \\ [\alpha]^- \end{array} & \begin{bmatrix} 1 & 0 \\ 0 & 1 \end{bmatrix} \end{array}$$

(resp. $\begin{bmatrix} 0 & 1 \\ 1 & 0 \end{bmatrix}$ if we use another denumeration).

Gathering up we have obtained (see Puttaswamaiah [1], Puttaswamaiah/Robinson [1], Kerber [3]):

7.47 If $D^1_{S_n,p}$ contains the submatrix

$$\begin{array}{cc} & \begin{array}{cc} F_S & F_S \otimes \overline{[1^n]} \end{array} \\ \begin{array}{c} [\alpha] \\ [\alpha'] \end{array} & \begin{bmatrix} a & b \\ b & a \end{bmatrix} \end{array},$$

then if

(i) $\alpha \neq \alpha'$, $F_S \not\sim F_S \otimes \overline{[1^n]}$, $D^1_{A_n,p}$ contains the submatrix

$$\begin{array}{cc} & F_S \downarrow A_n \\ [\alpha]\downarrow A_n & \left[\, a+b \,\right] \end{array}$$

(ii) <u>If $\alpha \neq \alpha'$, $F_S \sim F_S \otimes \overline{[1^n]}$, $F_S \downarrow A_n$ irreducible:</u>

$$\begin{array}{cc} & F_S \downarrow A_n \\ [\alpha]\downarrow A_n & \left[\, a \,\right] \end{array}$$

(iii) <u>$\alpha \neq \alpha'$, $F_S \downarrow A_n \leftrightarrow F_S^+ + F_S^-$:</u>

$$\begin{array}{ccc} & F_S^+ & F_S^- \\ [\alpha]\downarrow A_n & \left[\, a \right. & \left. a \,\right] \end{array}$$

(iv) <u>$\alpha = \alpha'$, $F_S \not\sim F_S \otimes \overline{[1^n]}$:</u>

$$\begin{array}{cc} & F_S \downarrow A_n \\ [\alpha]^+ & \\ [\alpha]^- & \end{array} \begin{bmatrix} a \\ a \end{bmatrix}$$

(v) <u>$\alpha = \alpha'$, $F_S \sim F_S \otimes \overline{[1^n]}$, $F_S \downarrow A_n$ irreducible:</u>

$$\begin{array}{cc} & F_S \downarrow A_n \\ [\alpha]^+ & \\ [\alpha]^- & \end{array} \begin{bmatrix} a/2 \\ a/2 \end{bmatrix}$$

(vi) <u>$\alpha = \alpha'$, $F_S \leftrightarrow F_S^+ + F_S^-$, $a = 0$ resp. $a = 1$:</u>

$$\begin{array}{c} \\ [\alpha]^+ \\ [\alpha]^- \end{array} \begin{array}{cc} F_S^+ & F_S^- \end{array} \begin{bmatrix} 0 & 0 \\ 0 & 0 \end{bmatrix} \quad \text{resp.} \quad \begin{bmatrix} 1 & 0 \\ 0 & 1 \end{bmatrix}.$$

As an example we give the decomposition matrix of A_9 which arises from the decomposition matrix of S_9 for $p = 3$ given in Kerber/Peel [1]:

7.48

$$D^1_{A_{9,3}} = \left[\begin{array}{ccccc|cc|c}
1 & & & & & & & \\
1 & 1 & & & & & & \\
0 & 1 & 1 & & & & & \\
0 & 1 & 0 & 1 & & & & \\
1 & 1 & 1 & 1 & 1 & & & \\
0 & 0 & 1 & 0 & 1 & & & \\
1 & 0 & 0 & 1 & 0 & & & \\
2 & 1 & 0 & 1 & 2 & & & \\
0 & 0 & 0 & 0 & 1 & & & \\
0 & 0 & 0 & 0 & 1 & & & \\
1 & 1 & 0 & 1 & 1 & & & \\
1 & 2 & 2 & 1 & 2 & & & \\
0 & 0 & 1 & 0 & 0 & & & \\
0 & 0 & 1 & 0 & 0 & & & \\
\hline
 & & & & & 1 & & \\
 & & & & & 1 & 1 & \\
 & & & & & & 1 & \\
\hline
 & & & & & & & 1
\end{array}\right]
\begin{array}{l}
[9] \\
[8,1] \\
[7,1^2] \\
[6,3] \\
[6,2,1] \\
[6,1^3] \\
[5,4] \\
[5,2^2] \\
[5,1^4]^+ \\
[5,1^4]^- \\
[4^2,1] \\
[4,3,2] \\
[3^3]^+ \\
[3^3]^- \\
[7,2] \\
[4,2^2,1] \\
[4,2,1^3] \\
[5,3,1]
\end{array}$$

(Understand, that the row labelled by $[\alpha]$ indicates the decomposition of $[\alpha]\downarrow A_n$.)

For further examples see Puttaswamaiah [1], Puttaswamaiah/Robinson [1], Kerber [3] and Kerber/Peel [1].

8. Generalized decomposition numbers of symmetric and alternating groups

If p is a prime number and G a finite group with ordinary irreducible characters ζ^i, Brauer characters φ^k of the irreducible p-modular representations, and decomposition numbers d^1_{ik} with respect to p we have for a p-regular element $g \in G$:

8.1 $$\zeta^i(g) = \sum_k d^1_{ik}\varphi^k(g) .$$

This can be generalized to general group elements g.

As is well known an element $g \in G$ is a product of a uniquely determined p-element x with a uniquely determined p-regular y which commutes with x:

8.2 $$g = xy = yx .$$

Let us call x the <u>p-component</u>, y the <u>p-regular component of</u> g. g runs through a complete system of representatives of the conjugacy classes of G if in 8.2 x runs through a complete system of the p-classes of G and y - while x is fixed - runs through a complete system of representatives of the p-regular classes of the centralizer $C_G(x)$ of x in G. Since $C_G(1) = G$, the following result of Brauer generalizes 8.1:

<u>8.3 If $x \in G$ is a p-element of order p^r and if $\hat{\varphi}^k$ are the Brauer</u>

characters of $C_G(x)$ with respect to p, then there exist algebraic integers d^x_{ik} in $Q(\varepsilon)$ (ε a primitive p^r-th root of unity) depending only on x and satisfying

$$\zeta^i(xy) = \sum_k d^x_{ik} \hat{\varphi}^k(y) \ ,$$

$y \in C_G(x)$, y p-regular.

(cf. Curtis/Reiner [1], § 90A)

If now D^x indicates the matrix of these algebraic integers

8.4 $$D^x := (d^x_{ik}) \ ,$$

it can be shown, that for an x' conjugate to x the matrix $D^{x'}$ arises from D^x by a permutation of the columns. Thus for an investigation of these algebraic integers d^x_{ik} we need only consider the matrices D^{x_j} for a complete system, say for $\{x_1 := 1, \ldots, x_u\}$, of representatives of the p-classes of G.

For such a fixed system of representatives we denote for short

8.5 $$D^\nu := (d^{x_\nu}_{ik}),\ 1 \le \nu \le u.$$

The matrix

8.6 $$D := (D^1, \ldots, D^u),$$

which is therefore uniquely determined up to a column permutation is called the generalized decomposition matrix of G with respect to p. Its entries are called the generalized decomposition numbers of G with respect to p.

Its first columns contain D^1, the decomposition matrix of G.

If y_j are the representatives of the p-regular classes of $C_G(x_\nu)$

we indicate as follows:

8.7 $$Z^\nu := (\zeta^i(x_\nu y_k)),\ \Phi^\nu := (\hat{\varphi}^i(y_k))$$

and hence the matrices

8.8 $$\hat{\Phi} := \dotplus_{\nu=1}^{u} \Phi^\nu\ ,\ Z := (Z^1,\ldots,Z^u)$$

satisfy the equation

8.9 $$Z = D\hat{\Phi}\ .$$

We would like to evaluate these matrices D of generalized decomposition numbers of the symmetric group.

At first we notice that since a matrix of Brauer characters is not singular:

8.10 $$D^\nu = Z^\nu(\Phi^\nu)^{-1}.$$

Z^ν is known from the character table, thus it remains to derive the matrices Φ^ν of the Brauer characters of centralizers of p-elements of S_n.

$\pi \in S_n$ is a p-element if and only if the lengths of all the cyclic factors of π are powers of p, this is implied by 1.11.

From 2.32 we obtain, that the centralizer of such a p-element is a direct product of symmetries of cyclic p-groups:

8.11 If $\pi \in S_n$ is a p-element of type $T\pi = (a_1,\ldots,a_n)$, then we have for the centralizer of π:

$$C_{S_n}(\pi) = \times_i \left(C_{p^i} \sim S_{a_{p^i}}\right)\ .$$

This implies (cf. also Osima [4]) that the wanted matrix Φ^π of the Brauer characters of the centralizer of the p-element π is a direct product of the matrices $\Phi_{a_{p^i}}$ of the Brauer characters of the $S_{a_{p^i}}$:

8.12 $$\Phi^\pi = \times_i \Phi_{a_{p^i}} .$$

Since Q and hence also GF(p) is a splitting field for S_n, the $\Phi_{a_{p^i}}$ and therefore Φ^π is rational integral. Hence (Osima [4]):

8.13 <u>The Brauer characters with respect to p of the centralizers of p-elements of symmetric groups are rational integral.</u>

Hence $(\Phi^\pi)^{-1}$ is a matrix with rational entries and therefore the same is valid for D^ν (see 8.10) since the ordinary irreducible characters of S_n are rational integral valued. And since the generalized decomposition numbers are algebraic integers in $Q(\varepsilon)$ (see 8.3), they are not only rational but even rational integral (Kerber [1], see also Osima [4]):

8.14 <u>The generalized decomposition numbers of symmetric groups are rational integral.</u>

For an example we evaluate $D_{6,2}$, the generalized decomposition matrix of S_6 for $p = 2$.

(i) The 2-classes of S_6 are represented by $\pi_1 = 1$, $\pi_2 = (12)$, $\pi_3 = (12)(34)$, $\pi_4 = (12)(34)(56)$, $\pi_5 = (1234)$, $\pi_6 = (1234)(56)$.

The centralizers of these elements are of the form

$C(\pi_1) = S_6$, $C(\pi_2) = C_2 \times S_4$, $C(\pi_3) = C_2 \sim S_2 \times S_2$,

$C(\pi_4) = C_2 \sim S_3$, $C(\pi_5) = C_4 \times S_2$, $C(\pi_6) = C_4 \times C_2$.

(ii)Hence we need only the matrices of Brauer characters of S_1, S_2, S_3 and S_4. These matrices are

$$\Phi_1 = \Phi_2 = (1) \qquad \text{and} \qquad \Phi_3 = \Phi_4 = \begin{bmatrix} 1 & 1 \\ 2 & -1 \end{bmatrix},$$

as can be evaluated easily with the known character tables and the decomposition matrices $D^1_{2,2}$, $D^1_{3,2}$ and $D^1_{4,2}$ (see section 7). Thus we obtain

$$\Phi^2 = \Phi_4 \times \Phi_1 = \begin{bmatrix} 1 & 1 \\ 2 & -1 \end{bmatrix}, \quad \Phi^3 = \Phi_2 \times \Phi_2 = (1) ,$$

$$\Phi^4 = \Phi_3 = \begin{bmatrix} 1 & 1 \\ 2 & -1 \end{bmatrix}, \quad \Phi^5 = \Phi_1 \times \Phi_2 = (1) = \Phi^6 .$$

(Φ^1 has been omitted since $D^1_{6,2}$ has been evaluated in section 7).

Because

$$\begin{bmatrix} 1 & 1 \\ 2 & -1 \end{bmatrix}^{-1} = \begin{bmatrix} 1/3 & 1/3 \\ 2/3 & -1/3 \end{bmatrix}$$

and

$$Z^2 = \begin{bmatrix} 1 & 1 \\ 3 & 0 \\ 3 & 0 \\ 2 & -1 \\ 1 & 1 \\ 0 & 0 \\ -1 & -1 \\ -2 & 1 \\ -3 & 0 \\ -3 & 0 \\ -1 & -1 \end{bmatrix} \qquad \text{and} \qquad Z^4 = \begin{bmatrix} 1 & 1 \\ -1 & -1 \\ 3 & 0 \\ -2 & 1 \\ -3 & 0 \\ 0 & 0 \\ 3 & 0 \\ 2 & -1 \\ -3 & 0 \\ 1 & 1 \\ -1 & -1 \end{bmatrix}$$

as can be read off from the character table of S_6, we get for the generalized decomposition matrix of S_6 with respect to p = 2:

$$
\text{8.15} \qquad D_{6,2} = \left[\begin{array}{ccc|c|cc|c|cc|c|c}
1 & & & & 1 & & 1 & 1 & & 1 & 1 \\
1 & 1 & & & 1 & 1 & 1 & -1 & & 1 & -1 \\
1 & 1 & 1 & & 1 & 1 & 1 & 1 & 1 & -1 & 1 \\
2 & 1 & 1 & & 0 & 1 & -2 & 0 & -1 & 0 & 0 \\
1 & 0 & 1 & & 1 & 0 & 1 & -1 & -1 & -1 & -1 \\
1 & 0 & 1 & & -1 & 0 & 1 & 1 & 1 & 1 & -1 \\
2 & 1 & 1 & & 0 & -1 & -2 & 0 & 1 & 0 & 0 \\
1 & 1 & 1 & & -1 & -1 & 1 & -1 & -1 & 1 & 1 \\
1 & 1 & & & -1 & -1 & 1 & 1 & & -1 & -1 \\
1 & & & & -1 & & 1 & -1 & & -1 & 1 \\
\hline
 & & & 1 & & & & & & &
\end{array}\right]
$$

(the columns belonging to $\Phi^3 = \Phi^5 = \Phi^6 = (1)$ agree with the concerning columns of the character table).

The generalized decomposition numbers of A_n can be evaluated similarly as long as the appropriate centralizers of p-elements are direct products of generalized alternating groups $C_{p^i} \sim A_{a_{p^i}}$ or have (1) as matrix of Brauer characters (such that the columns agrees with a column of the character table). In this way the generalized decomposition matrices of A_n with respect to p=3 have been calculated for $n \leq 7$ (Kerber [1]).

During these calculations we see that, otherwise than in the case of the symmetric group, the generalized decomposition numbers of A_n are not in general rational integers, e.g.

$$
\text{8.16} \qquad D_{A_3,3} := \left[\begin{array}{c|c|c}
1 & 1 & 1 \\
1 & (-1+i\sqrt{3})/2 & (-1-i\sqrt{3})/2 \\
1 & (-1-i\sqrt{3})/2 & (-1+i\sqrt{3})/2
\end{array}\right]
$$

is the generalized decomposition matrix of A_3 for p = 3. Thus

(Kerber [1]):

8.17 <u>The generalized decomposition numbers of alternating groups are not in general rational integral.</u>

Nevertheless (Osima [5]):

8.18 <u>The generalized decomposition numbers of alternating groups are rational integral for p = 2.</u>

Proof: The decomposition numbers of A_n are rational integral by definition, thus D^1 is a matrix over **Z**.
Hence it suffices to show that Φ^ν and the Z^ν are matrices over **Z** if $\nu>1$.
Thus it is enough to prove that the Brauer characters with respect to p=2 of centralizers of 2-elements $\pi\neq 1$ of A_n are rational integral and that this is valid also for the values of the ordinary irreducible characters of 2-singular elements of A_n.
The last statement is valid as can be seen from Frobenius' theorem 4.55, since permutations of splitting classes are obviously 2-regular. It remains to prove, that the values of the Brauer characters with respect to p=2 of centralizers of 2-elements in A_n are rational integral.
Thus the proof of the following lemma (Osima [5]) completes the proof of 8.18:

8.19 <u>If $\pi \neq 1$ is a 2-element of A_n, then the irreducible Brauer characters of $C_{S_n}(\pi)$ with respect to p=2 remain irreducible if they are restricted to $C_{A_n}(\pi)$ such that the values of the Brauer characters of $C_{A_n}(\pi)$ are rational integral as well (cf. 8.13).</u>

Proof: We shall show that there is a subgroup of $C_{A_n}(\pi)$ which has the same matrix of Brauer characters as $C_{S_n}(\pi)$.
Let π be an even 2-element of S_n such that $\pi \neq 1$ and hence

$$C_{S_n}(\pi) = \underset{i}{\times} (C_i \sim S_{a_i}) = \underset{j}{\times} (C_{2^j} \sim S_{a_{2^j}}) = \underset{j}{\times} (C^*_{2^j} S'_{a_{2^j}}) .$$

$\pi \neq 1$ implies that there is a $k \geq 1$ so that $a_{2^k} > 0$. Let us fix such a k. A subgroup of $C_{S_n}(\pi)$ with the same matrix of Brauer characters as $C_{S_n}(\pi)$ is

$$G := S'_{a_1} \times \ldots \times S'_{a_{2^{k-1}}} \times (C_{2^k} \sim S_{a_{2^k}}) \times S'_{a_{2^{k+1}}} \times \ldots .$$

Let us consider the subgroup

$$G^+ := G \cap A_n \leq C_{A_n}(\pi) .$$

We would like to show, that G^+ has the same irreducible Brauer characters as G and hence as $C_{S_n}(\pi)$, what implies that the statement 8.19 is valid.
From Clifford's theory of representations of groups with normal divisors we know, that a normal divisor N of index 2 in B has the same irreducible Brauer characters as B if no 2-regular conjugacy class of B splits into conjugacy classes of N.

Hence it suffices to show that no 2-regular class of G splits into two 2-regular classes of G^+.

If C^G is a 2-regular class of G it is a product

$$C^G = C^1C^2C^4\ldots$$

of 2-regular classes C^j of the direct factors $S'_{a_{2^j}}$ if $j \neq k$, and C^k is a 2-regular class of $C_{2^k} \sim S_{a_{2^k}}$.

From the results of section 3 we obtain that $C^k \subseteq S'_{a_{2^k}}$, such that all the factors C^i are contained in G^+ as well.

It is obvious, that the factors C^i don't split into G^+-classes if $i>1$. Hence it remains to show that C^1 does not split, which can be proved as follows.

Let ρ be any odd permutation out of S'_{a_1}; since $k>0$ there is an odd permutation $\sigma \in C_{2^k} \sim S_{a_{2^k}}$ and we have $\rho\sigma \in G^+$. Hence C^1 does not split.

This completes the proof of 8.19 and so also of 8.18.

q.e.d.

Hence (Osima [5]):

8.20 <u>If $\pi \neq 1$ is a 2-element of type $T\pi = (a_1,\ldots,a_n)$ of A_n, then $C_{S_n}(\pi)$ as well as $C_{A_n}(\pi)$ have</u>

$$\Phi_{a_1} \times \Phi_{a_2} \times \Phi_{a_4} \times \ldots$$

<u>as matrix of Brauer characters with respect to p=2 if $\Phi_{a_{2^j}}$ is the matrix of the 2-modular Brauer characters of $S_{a_{2^j}}$.</u>

Using this lemma M. Osima evaluated the generalized decomposition matrices of A_n with respect to p=2 for n=6,7,8,9 (Osima [5]), the matrices for $n \leq 7$ had already been known (Kerber [1]).
This shows, that for to prove that the generalized decomposition numbers of a finite group G with respect to p are rational integers it suffices to show that the values of the Brauer characters of centralizers of p-elements as well as the values of ordinary irreducible characters on p-singular elements are rational integral. We would like to check whether this is true for certain wreath products $G \sim S_n$.

8.21 If GF(p) is a splitting field for the centralizers of the p-elements $\neq 1$ of the finite group G and if the values of the ordinary irreducible characters with respect to p of $G \sim S_n$ on p-singular elements are rational integral, then the generalized decomposition numbers of $G \sim S_n$ with respect to p are rational integral.

Proof: Let $(f;\pi)$ be a p-element of $G \sim S_n$ and of type (a_{ik}). For its centralizer we have (cf. 3.25):

8.22 $$C_{G \sim S_n}(f;\pi) \simeq \underset{i,k}{\times}(C_{G \sim S_k}(f^j_{ik};\pi^j_{ik}) \sim S_{a_{ik}}) .$$

Hence it suffices to consider the Brauer characters of the factors

8.23 $$C_{G \sim S_k}(f^j_{ik};\pi^j_{ik}) \sim S_{a_{ik}} .$$

We would like to show that the values of their Brauer characters with respect to p are rational integral.

From the results of section 5 we obtain that it suffices to show that the p-modular representations of

8.24 $$C_{G\sim S_k}(f^i_{ik};\pi^j_{ik})$$

can be written over GF(p). But this is valid since 8.24 is an extension of the centralizer of a p-element of G with a cyclic p-group (cf. 3.19).

Thus GF(p) is a splitting field for the subgroups 8.24 and hence for the basis group of 8.23 as well. This implies (see section 5) that GF(p) is a splitting field for the groups 8.23, too, since it is a splitting field for symmetric groups $S_{a_{ik}}$ as well.

Hence GF(p) is a splitting field for $C_{G\sim S_n}(f;\pi)$ (see 8.22) such that the values of its Brauer characters are rational integral.

The decomposition numbers of $G\sim S_n$ are rational integral by definition, the values of the ordinary irreducible characters of $G\sim S_n$ on p-singular elements are rational integral by assumption, hence (cf. 8.10) the D^ν are matrices over $\mathbb{Z}$.

q.e.d.

8.11 and the results of section 5 imply that GF(p) is a splitting field for the centralizers of p-elements in S_n, thus as a special case of 8.21 we obtain:

<u>8.25</u> The generalized decomposition numbers of $S_m\sim S_n$ with respect to p are rational integral for all m, n and p.

As we have seen, the p-modular irreducible representations of the centralizers $C_{S_n}(\pi)$ of p-elements $\pi \neq 1$ such that $\pi \in A_n$ remain irreducible if they are restricted to $C_{A_n}(\pi)$ such that GF(p) is a splitting field for $C_{A_n}(\pi)$ as well. Hence also 8.18 is a special case of 8.21 ($G:=A_m$, $n=1$). For $G := \{1\}$ we obviously obtain 8.14.

Multiplying $D_{S_n,p}$ resp. $D_{A_n,p}$ with the transposed of its complex-conjugate we get the matrices

8.26
$$C_{S_n,p} := {}^t\overline{D_{S_n,p}} \cdot D_{S_n,p} = C^{\pi_1} \dotplus \ldots \dotplus C^{\pi_u} ,$$
$$C_{A_n,p} := {}^t\overline{D_{A_n,p}} \cdot D_{A_n,p} = C^{\rho_1} \dotplus \ldots \dotplus C^{\rho_v} ,$$

__the__ __generalized__ __Cartan__ __matrix__ of S_n resp. A_n, consisting of submatrices C^{π_i} resp. C^{ρ_j} along the leading diagonal which are the Cartan matrices of the centralizers of the p-elements π_i resp. ρ_j (cf. Curtis/Reiner [1], § 90A). For example

$$C_{S_6,2} = \begin{bmatrix} 16 & 8 & 8 \\ 8 & 6 & 4 \\ 8 & 4 & 6 \end{bmatrix} \dotplus [1] \dotplus \begin{bmatrix} 8 & 4 \\ 4 & 6 \end{bmatrix} \dotplus [16] \dotplus \begin{bmatrix} 8 & 4 \\ 4 & 6 \end{bmatrix} \dotplus [8] \dotplus [8].$$

The ordinary irreducible characters of the centralizer

$$C_{S_n}(\pi) = \mathop{\times}_{i=0}^{n} (C_{p^i} \sim S_{a_{p^i}})$$

of a p-element π of S_n of type $T\pi = (a_1,\ldots,a_n)$ are of course the products

$$\zeta = \zeta^0 \zeta^1 \ldots \zeta^n$$

of irreducible ordinary characters ζ^i of the direct factors

$$C_{p^i} \wr S_{a_{p^i}}$$

($\zeta^i := 1$, if $a_{p^i}=0$).

Since

$$C_{S_n}(\pi)/C_1 \wr S_{a_1} = \mathop{\times}_{i=1}^{n} (C_{p^i} \wr S_{a_{p^i}})$$

possesses only one p-block (cf. 5.31) we have (Osima [4]):

8.27 Two ordinary irreducible characters $\zeta=\zeta^0\ldots\zeta^n$ and $\zeta'=\zeta^{0'}\ldots\zeta^{n'}$ of $C_{S_n}(\pi)$ (π a p-element of type $T\pi=(a_1,\ldots,a_n)$) belong to the same p-block if and only if ζ^0 and $\zeta^{0'}$ belong to the same p-block of S_{a_1}, i.e. if they both have the same p-core.

According to this we denote the diagram of ζ^0 as the diagram of ζ such that now 8.27 reads as follows:

8.28 Two ordinary irreducible representations of the centralizer of a p-element of S_n belong to the same p-block if and only if their diagrams have the same p-core.

Two corollaries are (Kerber [1], Osima [4]):

8.29 If π is a p-element of type $(a_1,\ldots,a_n)$, then $C_{S_n}(\pi)$ has only one p-block if $a_1 \leq 1$ and $p \neq 2$ resp. if $a_1 \leq 2$ and $p=2$.

An application to the character table is:

8.30 <u>If B is the block of S_n which contains the identity representation $\overline{[n]}$, then $\zeta^{\alpha}(xy) = 0$, $\forall$ $[\alpha] \in B$ if $a_1 \leq 1$ (x a p-element of type $(a_1,\ldots,a_n)$) and $p \neq 2$ resp. if $a_1 \leq 2$ and $p=2$.</u>

These results follow from a result of Brauer (Brauer [2], main theorem), that every column of the generalized decomposition matrix of a finite group G contains nonvanishing entries only in the rows of a certain p-block (8.30 can be obtained from the Murnaghan-Nakayama-formula as well).

8.29 suggests that we conclude with a hint at an important result of Osima (Osima [4]) which establishes the connection between the p-blocks of S_n and the p-blocks of the centralizers of p-elements (see Brauer [2], (6A), also Curtis/Reiner [1], §§ 87, 90A):

8.31 <u>The block $\tilde{B}$ of $C_{S_n}(\pi)$ with p-core $[\tilde{\alpha}]$ determines the p-block B of S_n with the same p-core $[\tilde{\alpha}]$.</u>

Lecture Notes in Mathematics

Comprehensive leaflet on request

Vol. 38: R. Berger, R. Kiehl, E. Kunz und H.-J. Nastold, Differentialrechnung in der analytischen Geometrie IV, 134 Seiten. 1967 DM 12,–

Vol. 39: Séminaire de Probabilités I. II, 189 pages. 1967. DM 14,–

Vol. 40: J. Tits, Tabellen zu den einfachen Lie Gruppen und ihren Darstellungen. VI, 53 Seiten. 1967. DM 6,80

Vol. 41: A. Grothendieck, Local Cohomology. VI, 106 pages. 1967. DM 10,–

Vol. 42: J. F. Berglund and K. H. Hofmann, Compact Semitopological Semigroups and Weakly Almost Periodic Functions. VI, 160 pages. 1967. DM 12,–

Vol. 43: D. G. Quillen, Homotopical Algebra. VI, 157 pages. 1967. DM 14,–

Vol. 44: K. Urbanik, Lectures on Prediction Theory. IV, 50 pages. 1967. DM 5,80

Vol. 45: A. Wilansky, Topics in Functional Analysis. VI, 102 pages. 1967. DM 9,60

Vol. 46: P. E. Conner, Seminar on Periodic Maps. IV, 116 pages. 1967. DM 10,60

Vol. 47: Reports of the Midwest Category Seminar I. IV, 181 pages. 1967. DM 14,80

Vol. 48: G. de Rham, S. Maumary et M. A. Kervaire, Torsion et Type Simple d'Homotopie. IV, 101 pages. 1967. DM 9,60

Vol. 49: C. Faith, Lectures on Injective Modules and Quotient Rings. XVI, 140 pages. 1967. DM 12,80

Vol. 50: L. Zalcman, Analytic Capacity and Rational Approximation VI, 155 pages. 1968. DM 13,20

Vol. 51: Séminaire de Probabilités II. IV, 199 pages. 1968. DM 14,–

Vol. 52: D. J. Simms, Lie Groups and Quantum Mechanics. IV, 90 pages. 1968. DM 8,–

Vol. 53: J. Cerf, Sur les difféomorphismes de la sphère de dimension trois (Γ_4 = O). XII, 133 pages. 1968. DM 12,–

Vol. 54: G. Shimura, Automorphic Functions and Number Theory. VI, 69 pages. 1968. DM 8,–

Vol. 55: D. Gromoll, W. Klingenberg und W. Meyer, Riemannsche Geometrie im Großen. VI, 287 Seiten. 1968. DM 20,–

Vol. 56: K. Floret und J. Wloka, Einführung in die Theorie der lokalkonvexen Räume. VIII, 194 Seiten. 1968. DM 16,–

Vol. 57: F. Hirzebruch und K. H. Mayer, O (n)-Mannigfaltigkeiten, exotische Sphären und Singularitäten. IV, 132 Seiten. 1968. DM 10,80

Vol. 58: Kuramochi Boundaries of Riemann Surfaces. IV, 102 pages. 1968. DM 9,60

Vol. 59: K. Jänich, Differenzierbare G-Mannigfaltigkeiten. VI, 89 Seiten. 1968. DM 8,–

Vol. 60: Seminar on Differential Equations and Dynamical Systems. Edited by G. S. Jones. VI, 106 pages. 1968. DM 9,60

Vol. 61: Reports of the Midwest Category Seminar II. IV, 91 pages. 1968. DM 9,60

Vol. 62: Harish-Chandra, Automorphic Forms on Semisimple Lie Groups X, 138 pages. 1968. DM 14,–

Vol. 63: F. Albrecht, Topics in Control Theory. IV, 65 pages. 1968. DM 6,80

Vol. 64: H. Berens, Interpolationsmethoden zur Behandlung von Approximationsprozessen auf Banachräumen. VI, 90 Seiten. 1968. DM 8,–

Vol. 65: D. Kölzow, Differentiation von Maßen. XII, 102 Seiten. 1968. DM 8,–

Vol. 66: D. Ferus, Totale Absolutkrümmung in Differentialgeometrie und -topologie. VI, 85 Seiten. 1968. DM 8,–

Vol. 67: F. Kamber and P. Tondeur, Flat Manifolds. IV, 53 pages. 1968. DM 5,80

Vol. 68: N. Boboc et P. Mustată, Espaces harmoniques associés aux opérateurs différentiels linéaires du second ordre de type elliptique. VI, 95 pages. 1968. DM 8,60

Vol. 69: Seminar über Potentialtheorie. Herausgegeben von H. Bauer. VI, 180 Seiten. 1968. DM 14,80

Vol. 70: Proceedings of the Summer School in Logic. Edited by M. H. Löb. IV, 331 pages. 1968. DM 20,–

Vol. 71: Séminaire Pierre Lelong (Analyse), Année 1967 – 1968. VI, 190 pages. 1968. DM 14,

Vol. 72: The Syntax and Semantics of Infinitary Languages. Edited by J. Barwise. IV, 268 pages. 1968. DM 18,–

Vol. 73: P. E. Conner, Lectures on the Action of a Finite Group. IV, 123 pages. 1968. DM 10,–

Vol. 74: A. Fröhlich, Formal Groups. IV, 140 pages. 1968. DM 12,–

Vol. 75: G. Lumer, Algèbres de fonctions et espaces de Hardy. VI, 80 pages. 1968. DM 8,–

Vol. 76: R. G. Swan, Algebraic K-Theory. IV, 262 pages. 1968. DM 18,–

Vol. 77: P.-A. Meyer, Processus de Markov: la frontière de Martin. IV, 123 pages. 1968. DM 10,–

Vol. 78: H. Herrlich, Topologische Reflexionen und Coreflexionen. XVI, 166 Seiten. 1968. DM 12,–

Vol. 79: A. Grothendieck, Catégories Cofibrées Additives et Complexe Cotangent Relatif. IV, 167 pages. 1968. DM 12,–

Vol. 80: Seminar on Triples and Categorical Homology Theory. Edited by B. Eckmann. IV, 398 pages. 1969. DM 20,–

Vol. 81: J.-P. Eckmann et M. Guenin, Méthodes Algébriques en Mécanique Statistique. VI, 131 pages. 1969. DM 12,–

Vol. 82: J. Wloka, Grundräume und verallgemeinerte Funktionen. VIII, 131 Seiten. 1969. DM 12,–

Vol. 83: O. Zariski, An Introduction to the Theory of Algebraic Surfaces. IV, 100 pages. 1969. DM 8,–

Vol. 84: H. Lüneburg, Transitive Erweiterungen endlicher Permutationsgruppen. IV, 119 Seiten. 1969. DM 10,–

Vol. 85: P. Cartier et D. Foata, Problèmes combinatoires de commutation et réarrangements. IV, 88 pages. 1969. DM 8,–

Vol. 86: Category Theory, Homology Theory and their Applications I. Edited by P. Hilton. VI, 216 pages. 1969. DM 16,–

Vol. 87: M. Tierney, Categorical Constructions in Stable Homotopy Theory. IV, 65 pages. 1969. DM 6,–

Vol. 88: Séminaire de Probabilités III. IV, 229 pages. 1969. DM 18,–

Vol. 89: Probability and Information Theory. Edited by M. Behara, K. Krickeberg and J. Wolfowitz. IV, 256 pages. 1969. DM 18,–

Vol. 90: N. P. Bhatia and O. Hajek, Local Semi-Dynamical Systems. II, 157 pages. 1969. DM 14,–

Vol. 91: N. N. Janenko, Die Zwischenschrittmethode zur Lösung mehrdimensionaler Probleme der mathematischen Physik. VIII, 194 Seiten. 1969. DM 16,80

Vol. 92: Category Theory, Homology Theory and their Applications II. Edited by P. Hilton. V, 308 pages. 1969. DM 20,–

Vol. 93: K. R. Parthasarathy, Multipliers on Locally Compact Groups. III, 54 pages. 1969. DM 5,60

Vol. 94: M. Machover and J. Hirschfeld, Lectures on Non-Standard Analysis. VI, 79 pages. 1969. DM 6,–

Vol. 95: A. S. Troelstra, Principles of Intuitionism. II, 111 pages. 1969. DM 10,–

Vol. 96: H.-B. Brinkmann und D. Puppe, Abelsche und exakte Kategorien, Korrespondenzen. V, 141 Seiten. 1969. DM 10,–

Vol. 97: S. O. Chase and M. E. Sweedler, Hopf Algebras and Galois theory. II, 133 pages. 1969. DM 10,–

Vol. 98: M. Heins, Hardy Classes on Riemann Surfaces. III, 106 pages. 1969. DM 10,–

Vol. 99: Category Theory, Homology Theory and their Applications III. Edited by P. Hilton. IV, 489 pages. 1969. DM 24,–

Vol. 100: M. Artin and B. Mazur, Etale Homotopy. II, 196 Seiten. 1969. DM 12,–

Vol. 101: G. P. Szegö et G. Treccani, Semigruppi di Trasformazioni Multivoche. VI, 177 pages. 1969. DM 14,–

Vol. 102: F. Stummel, Rand- und Eigenwertaufgaben in Sobolewschen Räumen. VIII, 386 Seiten. 1969. DM 20,–

Vol. 103: Lectures in Modern Analysis and Applications I. Edited by C. T. Taam. VII, 162 pages. 1969. DM 12,–

Vol. 104: G. H. Pimbley, Jr., Eigenfunction Branches of Nonlinear Operators and their Bifurcations. II, 128 pages. 1969. DM 10,–

Vol. 105: R. Larsen, The Multiplier Problem. VII, 284 pages. 1969. DM 18,–

Vol. 106: Reports of the Midwest Category Seminar III. Edited by S. Mac Lane. III, 247 pages. 1969. DM 16,–

Vol. 107: A. Peyerimhoff, Lectures on Summability. III, 111 pages. 1969. DM 8,–

Vol. 108: Algebraic K-Theory and its Geometric Applications. Edited by R. M. F. Moss and C. B. Thomas. IV, 86 pages. 1969. DM 6,–

Vol. 109: Conference on the Numerical Solution of Differential Equations. Edited by J. Ll. Morris. VI, 275 pages. 1969. DM 18,–

Vol. 110: The Many Facets of Graph Theory. Edited by G. Chartrand and S. F. Kapoor. VIII, 290 pages. 1969. DM 18,–

Vol. 111: K. H. Mayer, Relationen zwischen charakteristischen Zahlen. III, 99 Seiten. 1969. DM 8,–

Vol. 112: Colloquium on Methods of Optimization. Edited by N. N. Moiseev. IV, 293 pages. 1970. DM 18,–

Vol. 113: R. Wille, Kongruenzklassengeometrien. III, 99 Seiten. 1970. DM 8,–

Vol. 114: H. Jacquet and R. P. Langlands, Automorphic Forms on GL (2). VII, 548 pages. 1970. DM 24,–

Vol. 115: K. H. Roggenkamp and V. Huber-Dyson, Lattices over Orders I. XIX, 290 pages. 1970. DM 18,–

Vol. 116: Séminaire Pierre Lelong (Analyse) Année 1969. IV, 195 pages. 1970. DM 14,–

Vol. 117: Y. Meyer, Nombres de Pisot, Nombres de Salem et Analyse Harmonique. 63 pages. 1970. DM 6,–

Vol. 118: Proceedings of the 15th Scandinavian Congress, Oslo 1968. Edited by K. E. Aubert and W. Ljunggren. IV, 162 pages. 1970. DM 12,–

Vol. 119: M. Raynaud, Faisceaux amples sur les schémas en groupes et les espaces homogènes. III, 219 pages. 1970. DM 14,–

Vol. 120: D. Siefkes, Büchi's Monadic Second Order Successor Arithmetic. XII, 130 Seiten. 1970. DM 12,–

Vol. 121: H. S. Bear, Lectures on Gleason Parts. III, 47 pages. 1970. DM 6,–

Vol. 122: H. Zieschang, E. Vogt und H.-D. Coldewey, Flächen und ebene diskontinuierliche Gruppen. VIII, 203 Seiten. 1970. DM 16,–

Vol. 123: A. V. Jategaonkar, Left Principal Ideal Rings. VI, 145 pages. 1970. DM 12,–

Vol. 124: Séminare de Probabilités IV. Edited by P. A. Meyer. IV, 282 pages. 1970. DM 20,–

Vol. 125: Symposium on Automatic Demonstration. V, 310 pages. 1970. DM 20,–

Vol. 126: P. Schapira, Théorie des Hyperfonctions. XI, 157 pages. 1970. DM 14,–

Vol. 127: I. Stewart, Lie Algebras. IV, 97 pages. 1970. DM 10,–

Vol. 128: M. Takesaki, Tomita's Theory of Modular Hilbert Algebras and its Applications. II, 123 pages. 1970. DM 10,–

Vol. 129: K. H. Hofmann, The Duality of Compact Semigroups and C^*-Bigebras. XII, 142 pages. 1970. DM 14,–

Vol. 130: F. Lorenz, Quadratische Formen über Körpern. II, 77 Seiten. 1970. DM 8,–

Vol. 131: A Borel et al., Seminar on Algebraic Groups and Related Finite Groups. VII, 321 pages. 1970. DM 22,–

Vol. 132: Symposium on Optimization. III, 348 pages. 1970. DM 22,–

Vol. 133: F. Topsøe, Topology and Measure. XIV, 79 pages. 1970. DM 8,–

Vol. 134: L. Smith, Lectures on the Eilenberg-Moore Spectral Sequence. VII, 142 pages. 1970. DM 14,–

Vol. 135: W. Stoll, Value Distribution of Holomorphic Maps into Compact Complex Manifolds. II, 267 pages. 1970. DM 18,–

Vol. 136: M. Karoubi et al., Séminaire Heidelberg-Saarbrücken-Strasbourg sur la K-Théorie. IV, 264 pages. 1970. DM 18,–

Vol. 137: Reports of the Midwest Category Seminar IV. Edited by S. MacLane. III, 139 pages. 1970. DM 12,–

Vol. 138: D. Foata et M. Schützenberger, Théorie Géométrique des Polynômes Eulériens. V, 94 pages. 1970. DM 10,–

Vol. 139: A. Badrikian, Séminaire sur les Fonctions Aléatoires Linéaires et les Mesures Cylindriques. VII, 221 pages. 1970. DM 18,–

Vol. 140: Lectures in Modern Analysis and Applications II. Edited by C. T. Taam. VI, 119 pages. 1970. DM 10,–

Vol. 141: G. Jameson, Ordered Linear Spaces. XV, 194 pages. 1970. DM 16,–

Vol. 142: K. W. Roggenkamp, Lattices over Orders II. V, 388 pages. 1970. DM 22,–

Vol. 143: K. W. Gruenberg, Cohomological Topics in Group Theory. XIV, 275 pages. 1970. DM 20,–

Vol. 144: Seminar on Differential Equations and Dynamical Systems, II. Edited by J. A. Yorke. VIII, 268 pages. 1970. DM 20,–

Vol. 145: E. J. Dubuc, Kan Extensions in Enriched Category Theory. XVI, 173 pages. 1970. DM 16,–

Vol. 146: A. B. Altman and S. Kleiman, Introduction to Grothendieck Duality Theory. II, 192 pages. 1970. DM 18,–

Vol. 147: D. E. Dobbs, Cech Cohomological Dimensions for Commutative Rings. VI, 176 pages. 1970. DM 16,–

Vol. 148: R. Azencott, Espaces de Poisson des Groupes Localement Compacts. IX, 141 pages. 1970. DM 14,–

Vol. 149: R. G. Swan and E. G. Evans, K-Theory of Finite Groups and Orders. IV, 237 pages. 1970. DM 20,–

Vol. 150: Heyer, Dualität lokalkompakter Gruppen. XIII, 372 Seiten. 1970. DM 20,–

Vol. 151: M. Demazure et A. Grothendieck, Schémas en Groupes I. (SGA 3). XV, 562 pages. 1970. DM 24,–

Vol. 152: M. Demazure et A. Grothendieck, Schémas en Groupes II. (SGA 3). IX, 654 pages. 1970. DM 24,–

Vol. 153: M. Demazure et A. Grothendieck, Schémas en Groupes III. (SGA 3). VIII, 529 pages. 1970. DM 24,–

Vol. 154: A. Lascoux et M. Berger, Variétés Kähleriennes Compactes. VII, 83 pages. 1970. DM 8,–

Vol. 155: Several Complex Variables I, Maryland 1970. Edited by J. Horváth. IV, 214 pages. 1970. DM 18,–

Vol. 156: R. Hartshorne, Ample Subvarieties of Algebraic Varieties. XIV, 256 pages. 1970. DM 20,–

Vol. 157: T. tom Dieck, K. H. Kamps und D. Puppe, Homotopietheorie. VI, 265 Seiten. 1970. DM 20,–

Vol. 158: T. G. Ostrom, Finite Translation Planes. IV. 112 pages. 1970. DM 10,–

Vol. 159: R. Ansorge und R. Hass. Konvergenz von Differenzenverfahren für lineare und nichtlineare Anfangswertaufgaben. VIII, 145 Seiten. 1970. DM 14,–

Vol. 160: L. Sucheston, Constributions to Ergodic Theory and Probability. VII, 277 pages. 1970. DM 20,–

Vol. 161: J. Stasheff, H-Spaces from a Homotopy Point of View. VI, 95 pages. 1970. DM 10,–

Vol. 162: Harish-Chandra and van Dijk, Harmonic Analysis on Reductive p-adic Groups. IV, 125 pages. 1970. DM 12,–

Vol. 163: P. Deligne, Equations Différentielles à Points Singuliers Reguliers. III, 133 pages. 1970. DM 12,–

Vol. 164: J. P. Ferrier, Seminaire sur les Algebres Complètes. II, 69 pages. 1970. DM 8,–

Vol. 165: J. M. Cohen, Stable Homotopy. V, 194 pages. 1970. DM 16,–

Vol. 166: A. J. Silberger, PGL_2 over the p-adics: its Representations, Spherical Functions, and Fourier Analysis. VII, 202 pages. 1970. DM 18,–

Vol. 167: Lavrentiev, Romanov and Vasiliev, Multidimensional Inverse Problems for Differential Equations. V, 59 pages. 1970. DM 10,–

Vol. 168: F. P. Peterson, The Steenrod Algebra and its Applications: A conference to Celebrate N. E. Steenrod's Sixtieth Birthday. VII, 317 pages. 1970. DM 22,–

Vol. 169: M. Raynaud, Anneaux Locaux Henséliens. V, 129 pages. 1970. DM 12,–

Vol. 170: Lectures in Modern Analysis and Applications III. Edited by C. T. Taam. VI, 213 pages. 1970. DM 18,–

Vol. 171: Set-Valued Mappings, Selections and Topological Properties of 2^X. Edited by W. M. Fleischman. X, 110 pages. 1970. DM 12,–

Vol. 172: Y.-T. Siu and G. Trautmann, Gap-Sheaves and Extension of Coherent Analytic Subsheaves. V, 172 pages. 1971. DM 16,–

Vol. 173: J. N. Mordeson and B. Vinograde, Structure of Arbitrary Purely Inseparable Extension Fields. IV, 138 pages. 1970. DM 14,–

Vol. 174: B. Iversen, Linear Determinants with Applications to the Picard Scheme of a Family of Algebraic Curves. VI, 69 pages. 1970. DM 8,–

Vol. 175: M. Brelot, On Topologies and Boundaries in Potential Theory. VI, 176 pages. 1971. DM 18,–

Vol. 176: H. Popp, Fundamentalgruppen algebraischer Mannigfaltigkeiten. IV, 154 Seiten. 1970. DM 16,–

Vol. 177: J. Lambek, Torsion Theories, Additive Semantics and Rings of Quotients. VI, 94 pages. 1971. DM 12,–

Vol. 178: Th. Bröcker und T. tom Dieck, Kobordismentheorie. XVI, 191 Seiten. 1970. DM 16,–

Vol. 179: Seminaire Bourbaki – vol. 1968/69. Exposés 347-363. IV. 295 pages. 1971. DM 22,–

Vol. 180: Séminaire Bourbaki – vol. 1969/70. Exposés 364-381. IV, 310 pages. 1971. DM 22,–

Vol. 181: F. DeMeyer and E. Ingraham, Separable Algebras over Commutative Rings. V, 157 pages. 1971. DM 16,–